What Should Individuals Do about Climate Change?

Climate change is a pressing problem. Does each of us have a moral responsibility to help tackle it? In this volume, Marion Hourdequin and Dan Shahar debate the timely issue of individual behavior and climate change, examining what it takes to live morally in a warming world.

Hourdequin argues there are important reasons for people to translate their concerns about climate change into actions in their personal lives. This includes attending to the many ways a single individual can help catalyze systemic change through choices about voting and political participation, food and clothing, energy use, travel, and so on. Shahar disagrees because he endorses moral specialization and division of labor in a world filled with many problems. He argues we should not expect everyone to take action on every serious issue: rather, it is acceptable and even desirable for people to focus on certain issues and decline to act on others – including climate change. The two authors take turns responding to each other and then defending their ultimate conclusions. This volume is sure to draw attention to the question of "individual choice" in climate change debates and to help clarify some of the best thinking on this issue.

Marion Hourdequin is a Professor of Philosophy at Colorado College whose work focuses on environmental philosophy, climate ethics, and relational ethics. She is the author of *Environmental Ethics: From Theory to Practice* (2nd edition, 2024) and co-editor, with David Havlick, of *Restoring Layered Landscapes* (2016). She served as President of the International Society for Environmental Ethics from 2022 to 2024, and she is an Associate Editor for the journal *Environmental Ethics*.

Dan C. Shahar is MBA Program Director and a Teaching Assistant Professor at West Virginia University. His research explores the moral and political dimensions of humanity's relationship with the natural world. He is the author of *Why It's OK to Eat Meat* (Routledge, 2021), co-editor (with David Schmidtz) of *Environmental Ethics: What Really Matters, What Really Works* (3rd edition, 2018), and author of over a dozen journal articles and book chapters.

Allen Thompson is Associate Professor of Ethics and Environmental Philosophy at Oregon State University.

Little Debates about Big Questions

Tyron Goldschmidt
Fellow of the Rutgers Center for Philosophy of Religion, USA

Dustin Crummett
University of Washington, Tacoma, USA

About the series:

Philosophy asks questions about the fundamental nature of reality, our place in the world, and what we should do. Some of these questions are perennial: for example, *Do we have will? What is morality?* Some are much newer: for example, *How far should free speech on campus extend? Are race, sex and gender social constructs?* But all of these are among the big questions in philosophy and they remain controversial.

Each book in the *Little Debates about Big Questions* series features two professors on opposite sides of a big question. Each author presents their own side, and the authors then exchange objections and replies. Short, lively, and accessible, these debates showcase diverse and deep answers. Pedagogical features include standard form arguments, section summaries, bolded key terms and principles, glossaries, and annotated reading lists.

The debate format is an ideal way to learn about controversial topics. Whereas the usual essay or book risks overlooking objections against its own proposition or misrepresenting the opposite side, in a debate each side can make their case at equal length, and then present objections the other side must consider. Debates have a more conversational and fun style too, and we selected particularly talented philosophers—in substance and style—for these kinds of encounters.

Debates can be combative—sometimes even descending into anger and animosity. But debates can also be cooperative. While our authors disagree strongly, they work together to help each other and the reader get clearer on the ideas, arguments, and objections.

This is intellectual progress, and a much-needed model for civil and constructive disagreement.

The substance and style of the debates will captivate interested readers new to the questions. But there's enough to interest experts too. The debates will be especially useful for courses in philosophy and related subjects—whether as primary or secondary readings— and a few debates can be combined to make up the reading for an entire course.

We thank the authors for their help in constructing this series. We are honored to showcase their work. They are all preeminent scholars or rising-stars in their fields, and through these debates they share what's been discovered with a wider audience. This is a paradigm for public philosophy, and will impress upon students, scholars, and other interested readers the enduring importance of debating the big questions.

Published Titles:

Should Wealth Be Redistributed?: A Debate
By Steven McMullen and James R. Otteson

Do We Have Free Will?: A Debate
By Robert Kane and Carolina Sartorio

Is There a God?: A Debate
By Kenneth L. Pearce and Graham Oppy

Is Political Authority an Illusion?: A Debate
By Michael Huemer and Daniel Layman

Selected Forthcoming Titles:

Should We Want to Live Forever?: A Debate
By Stephen Cave and John Martin Fischer

Consequentialism or Virtue Ethics?: A Debate
By Jorge L.A. Garcia and Alastair Norcross

For more information about this series, please visit: https://www.routledge.com/Little-Debates-about-Big-Questions/book-series/LDABQ

What Should Individuals Do about Climate Change?

A Debate

Marion Hourdequin and
Dan C. Shahar

Routledge
Taylor & Francis Group

NEW YORK AND LONDON

Designed cover image: Getty Images / Shelyna Long

First published 2025
by Routledge
605 Third Avenue, New York, NY 10158

and by Routledge
4 Park Square, Milton Park, Abingdon, Oxon, OX14 4RN

Routledge is an imprint of the Taylor & Francis Group, an informa business

© 2025 Marion Hourdequin and Dan C. Shahar

ISBN: 978-0-367-70455-1 (hbk)
ISBN: 978-0-367-70454-4 (pbk)
ISBN: 978-1-003-14643-8 (ebk)

DOI: 10.4324/9781003146438

Typeset in Sabon
by Newgen Publishing UK

Contents

Foreword   ix
ALLEN THOMPSON

Opening Statements   1

1   Why Individual Emissions Matter   3
    MARION HOURDEQUIN

2   Facing Our Limits   47
    DAN C. SHAHAR

Replies   91

3   Reply to Shahar: The Pervasiveness of Climate
    Change Provides Reasons and Opportunities to Act   93
    MARION HOURDEQUIN

4   Reply to Hourdequin: Defending Specialization
    and Division of Moral Labor   114
    DAN C. SHAHAR

Closing Statements   133

5   Individual Actions Matter for Climate Change   135
    MARION HOURDEQUIN

6   **Embracing Diversity among Altruists**   149
    DAN C. SHAHAR

    Notes   167
    Further Reading   175
    Glossary   178
    Bibliography   187
    Index   204

Foreword

Allen Thompson

Anthropogenic climate change is one of the most significant challenges facing humanity today, arguably *the most* significant challenge. We are now well past debating if global climate conditions are changing, if anthropogenic greenhouse gas emissions are the primary driver, or if we need international cooperation to mitigate the significant threats that climate change poses. Overwhelming scientific consensus supports that climate change is real, is primarily driven by carbon emissions generated by burning fossil fuels for energy, and currently poses and will continue to pose significant dangers and, if unabated, even greater dangers to human beings, future generations, extant species, and other features of the world that we value. It would be a colossal waste of time to debate any of these issues and time is of the essence.

The United Nations Framework Convention on Climate Change (UNFCCC) was established at the 1992 UN Conference on Environment and Development in Rio de Janeiro (i.e., the "Earth Summit") to organize international cooperation toward the goal of preventing "dangerous anthropogenic interference with the climate system" (Article 2). To avoid such danger, the UNFCCC aims to limit the increase of the global mean surface temperature to no more than +2° Celsius from the preindustrial average.[1] However, despite more than thirty years of big meetings and successive agreements (e.g., the 1997 Kyoto Protocol and the 2016 Paris Agreement) both annual anthropogenic global greenhouse gas emissions and the cumulative amount of anthropogenic greenhouse gases in the atmosphere continue to grow. But to have a decent change of limiting global warming to +2° Celsius, annual global greenhouse gas emissions must be cut in half by 2030 (only six years from date of writing) and must reach net zero by 2050. Not only is time short but real reductions in global emissions are just not happening. If (or, as is almost certain, when) we pass the +2° Celsius target, the world won't end. But conditions for people, animals, and historic ecosystems will be much more dangerous indeed, and increasingly so as the mean temperature grows.

Many of the undesirable impacts of climate change will occur in specific locations and many of the associated harms will befall particular people. Still, stepping back, we recognize the scale of the problem is global and any adequate, long-term response will require agreement and follow-through on coordinated actions by agents that are nation states and multi-national organizations. In short, the problem is enormous. But the problem is anthropogenic; it is clear what "we" should do, collectively: immediately and significantly reduce global greenhouse gas emissions. Since we are falling short of our collective responsibility, however, it is *far less* clear what each of us, as individual agents, should do in response climate change.

Regarding this question, the reader will find, reasonable people can disagree. In their engaging exchange, Marion Hourdequin and Dan Shahar don't take clearly opposing, "pro" and "con" positions on a single question so much as articulate and further elucidate distinct perspectives about the kind of ethical problems posed by climate change and what sorts of reasons individuals have (or don't have) for taking responsive action. The authors present different views about how social norms, social structures, and individual incentives could operate or be understood to mobilize meaningful behaviors. Their discussion reveals, I believe, significantly different conceptions of morality, how it connects us with others in society, and how it reflects different ways of understanding our personal identity. To anticipate the authors' fuller discussion, I will briefly outline how their perspectives differ along these dimensions. First, however, let's consider the most common framing of the question at hand, which is that we should focus on a moral accounting of our personal greenhouse gas emissions.

The question "What should individuals do about climate change?" is often understood as a question about a moral obligation to make lifestyle choices that will reduce one's personal greenhouse gas emissions. This is because we know different lifestyle choices can result in more (or less) emissions, we know climate change is driven by greenhouse gases, and that a greater amount of greenhouse gases emissions will deliver an increasing risk of greater climate harms.

So, an individual may wonder: do I have a moral duty, for example, to fly less? To buy an electric or hybrid car? To lower my thermostat in the winter and my air-conditioning in the summer? Or, the example that philosopher Walter Sinnott-Armstrong considers: do I have a moral obligation to abstain from Sunday pleasure drives in my ordinary, gasoline-powered automobile?

The puzzle can be parsed as the problem of inconsequentialism: admittedly, the emissions from any particular action undertaken by an individual are barely a drop in the ocean. On one hand, emissions from my Sunday pleasure drive alone do not cause climate change and so do not cause any of the harms attributable to climate change. On the other hand, I cannot prevent any harms caused by climate change by simply *not* taking my Sunday drive. Further, it looks like I could only have a moral obligation to refrain from some action if the action caused harm to others, or I could avoid wrongful harms to others by not so acting. Yet neither causal condition holds regarding the emissions attributable to my Sunday pleasure drive (or, for that matter, any of my other individual lifestyle choices): climate change and the associated harms will happen whether I take my drive or not. Thus, as an individual, I do not have any moral obligation based on carbon emissions to refrain from my Sunday drive.

At the same time, if we believe there are moral reasons to avoid harmful climate impacts, then this conclusion can seem problematic: we are collectively morally responsible to reduce aggregate emissions and many of these emissions come from the actions of individuals, but still no individual has a moral obligation to reduce their unnecessary personal emissions. Suitably qualified, Shahar will agree with this conclusion. Hourdequin, however, believes the conclusion is too quick. She also prefers talk about moral reasons an individual has for taking climate action, rather than strict obligations.

Fortunately, neither Hourdequin nor Shahar believe reducing personal greenhouse gas emissions should be the exclusive focus for questions about climate action for individuals. Both recognize people have "broad latitude" in deciding how they could act to make the world a better place, including various responses to climate change. *Aggregate* emissions reductions, of course, are essential to avoiding the worst climate outcomes. But people can respond well to climate change by political engagement, for example, or educating others, and so on, where the measure of moral value is not captured only by counting personal carbon emissions.

If curtailing personal emissions, even over a lifetime, isn't the only concern, then how should we understand what's at stake with the question of "what should I do, personally, about climate change?" On this the authors have well-developed but diverging views. You will find in their initial position statements, thoughtful responses to one another, and closing summaries, Hourdequin and Shahar maintain alternative conceptions about the kind of threat posed by climate change and they provide an informed exchange about the normative structure of answers to the question of what should be done about it, which ultimately reveals different conceptions of who we are as individual persons. The authors *agree* that climate change is a serious problem and there are many ways people can appropriately respond; they *disagree* about the threat posed by climate change, the relevant moral normativity, and I think, ultimately, about ethics and the constitution of a person's ethical agency. In the end, both positions are reasonable. There's no crucial error to be found here, on one side or the other. Taking a side in this debate, it seems, is to hold one of two competing ways to think about who we are and the best ways to respond to problems posed by climate change.

Consider how the authors differ in their views about the problem of climate change. Shahar thinks climate change is *one of several* serious problems facing humanity. If we start by considering the life prospects and standards of living as experienced by most people prior to the industrial revolution, then in terms of the probability of avoiding hunger, disease, violence, and extreme poverty, today most of us live as the luckiest of all. The great, general comparative prosperity has been provided by economic growth, which has been driven in large part by "burning fossil fuels for power and the transformation of nature for agriculture and settlement," phenomena that also have the unintended side effect of changing the climate. Despite the acknowledged climate-related harms and damages, according to Shahar, "[m]ainstream [economic] assessments project that even without aggressive measures to combat climate change, future people will live better, more prosperous lives on average than people today." But even if, despite unmitigated climate change, the *average* level of prosperity continues to grow, many particular people still face extreme poverty, significant oppression, hunger and disease. Such conditions give an altruist many alternatives for moral engagement.

Shahar argues that although we morally should be altruistic, the efficiency created by a division of labor makes it permissible for altruists to specialize, and thus ("in a world where large-scale public coordination is absent") there is no specific moral duty for altruists to "take costly unilateral action on climate change" including efforts to reduce their personal emissions, or donate to environmental nonprofits, or become political climate advocates, and so on. On Shahar's view of the problem, although one morally should be an altruist, one can remain an ethical person without working on climate change specifically.

Hourdequin sees things differently. While some harmful impacts of climate change are occurring already, "[w]ithout swift and concerted action, the harmful effects of climate change will continue to accelerate, leaving future generations with a damaged and depleted planet." From this perspective, climate change is not simply one of several important problems; it is not only an issue that may be of some altruists' special interest. Instead, the environmental alterations driven by anthropogenic climate change constitutes radical and unprecedented environmental *degradation* and are thus detrimental to the very prospects of human flourishing (even despite any increase in overall "economic prosperity"). Climate change is a phenomenon that does and will increasingly exacerbate other problems, such as political oppression, poverty, hunger and food insecurity, poor health and disease. For Hourdequin, the moral call for individuals to address climate change is implicated by a moral concern with almost any, or perhaps all, of the other major problems facing humanity. Thus, she maintains that individuals have significant reasons to act in response to a changing climate, including *reasons of integrity* and *reasons of relationality*.

Integrity is a virtue concerning "the relationship between values and actions" that "presses us to live by our values, and to harmonize [all] the values we have." Those who care about climate change, or any of the problems exacerbated by climate change, have reasons to act accordingly, including but not limited to personal emissions reductions.[2] In addition to this *intra*personal dimension, integrity has an *inter*personal dimension; integrating "concern about climate change with other values" moves one toward considering how to act not only for one's self, family, and friends, but also for other "climate-affected communities and ecosystems around the world" and so how to "effect constructive change" at "multiple levels," to "transform economies, institutions, and cultures in a way that enables current and future generations to flourish."

Hourdequin also holds a conception of *"persons as relational and of ethics as involving the character and quality of human relations with one another and the broader world."* From a contrasting, persons-as-atomistic perspective, we each have prudential reasons to advance our own interests, while moral reasons are anchored in protecting other autonomous persons from harm, where harm is understood as diminishment in one's ability to pursue their own interests. Increasing the general welfare, then, is just growing the aggregate of individual persons' preference satisfaction. But from a persons-as-relational perspective, we each have reasons to advance strong and healthy community relations because "interdependency is fundamental to the human condition." On this view, the common welfare is a measure of our collective well-being as a community, together. Climate change threatens our common, collective good and so we each also have moral reasons of relationality to act in solidarity protecting the community, which is fundamental to protecting the individuals thereof. We may see ourselves as belonging to a relatively *local community* with shared culture, or a wider, multi-generational *human family*, or as members of what Leopold called the *"land community,"* that is, the community of deeply interdependent humans, non-human animals, plants, soils, and water, etc. Or we may see ourselves as members of all these communities. Thus, *"all of our actions matter ethically,"* Hourdequin maintains, "insofar as they shape the character and quality of these relations."

Shahar, however, is concerned about moralizing the climate actions of individuals. Climate-impacting behaviors aren't bad in themselves – consider, again, a Sunday pleasure drive. Indeed, further, each of us is implicated in climate change by a myriad of "normal, everyday activities" that bring *benefit* to ourselves and, arguably, to others, e.g., "traveling, working, eating, shopping, and powering our homes, among countless other things," which don't warrant the specter of moral condemnation. So, although we are complicit in creating the problem, Shahar advocates for structural solutions and personal incentives to reduce excess and increase coordination toward the achievement of more desirable outcomes. For example, he mentions replacing coal with natural gas in power plants, while increasing nuclear and renewable sources of energy. Carbon taxes would make high-emitting choices more costly, lowering their frequency, while subsidies for lower-emitting choices would increase their frequency.

Even supposing we don't presently have moral obligations to reduce personal emissions, many in the environmental community aspire to develop new social norms which, indeed, would provide the grounds for new moral obligations for individuals to reduce their personal emissions. But Shahar sees social norms relevant to climate change mitigation as prone to inefficiency, in part because they constrain the flexibility in particular circumstances that is made possible by prioritizing individual liberty and personal incentives; from his perspective, social norms often amount to objectionable external constraints. Hourdequin, by contrast, sees work towards developing new, climate-relevant social norms as the process of building a responsive culture of caring and thus, from her relational view of persons, as an internal and authentic expression of who we are and, perhaps, who we want to be.

Who are we? What kind of problem is climate change? What, ethically, does responding to climate change call on me to do? These are among the most vital questions of our day. Above, I wrote that "[t]aking a side in this debate, it seems, is to hold one of two competing ways to think about who we are and the best ways to respond to problems posed by climate change." But F. Scott Fitzgerald says in "The Crack Up" (1936) that the "test of a first-rate intelligence is the ability to hold two opposed ideas in the mind at the same time, and still retain the ability to function." We have good reasons to believe that our very persons are both rationally autonomous beings at liberty and motivated to pursue our own interests and, at the same time, constituted by the communities we belong to in a network of relations that make us who we are. So, no, we don't have to take sides, we don't have to reject either one of these opposing ideas. But we do have to use our first-rate intelligence to function well toward making the world a better place. Who do we *want* to be? I hope we want to be good planetary stewards. If that's what we really want, then I'm confident considering the insights and ideas debated in this book can help us find our way.

Opening Statements

Chapter 1

Opening Statement
Why Individual Emissions Matter

Marion Hourdequin

Contents

1. Introduction   3
2. A Provocation: Encountering an Argument Against
   Individual Emissions Reductions   5
3. Why Do Personal Emissions Matter? Reasons of Integrity   17
   3.1. From Hypocrisy to Integrity   17
   3.2. What Is Integrity, and Why Does it Matter?   22
   3.3. Integrity-Based Reasons for Personal Emissions
        Reductions   23
   3.4. Is Climate Change the Government's Responsibility
        (Only)?   26
4. Why Do Personal Emissions Matter? Reasons of Relationality   27
   4.1. An Individualist Analysis of Climate Change: The Tragedy
        of the Commons   32
   4.2. A Relational Analysis of Climate Change:
        Interconnectedness and Relational Reasons   36
   4.3. A Broad View of Relational Reasons and its Connection to
        Climate Change   39
5. Polycentrism, Power, and Possibilities   41
   5.1. Polycentrism and Individual Action   43

1. Introduction

The effects of global climate change are now clearly visible throughout the world: sea levels are rising, coral reefs are dying, and storms have grown more severe. Human societies are experiencing significant challenges as coastlines erode in some areas, and droughts destroy farmland in others. Numerous species are threatened with extinction as ecological conditions change faster than

DOI: 10.4324/9781003146438-2

they can adapt. Without swift and concerted action, the harmful effects of climate change will continue to accelerate, leaving future generations with a damaged and depleted planet.

Under the UN Framework Convention on Climate Change (United Nations 1992), international negotiations to cut emissions and slow global warming have been ongoing for 30 years (for current activities and updates, see https://unfccc.int). However, progress has been slow, and nations around the world have not yet demonstrated commitments to emissions cuts sufficient to achieve the target of keeping global warming under 2 degrees Celsius. As national and international level efforts fall short, it seems clear that action at many levels is required. Cities and local governments have networked to develop plans and share strategies, and state and regional initiatives in the U.S. – such as California's low-carbon and fuel efficiency standards or the Regional Greenhouse Gas Initiative in the Northeast (www.rggi.org) – have the potential to shift markets, incentivize production of more fuel-efficient cars, and drive transitions to renewable energy (for related discussion, see Bernstein and Hoffman 2018; Hultman et al. 2020; Axsen and Wolinetz 2023). These local and regional initiatives are important elements of a "polycentric," multi-pronged approach to climate mitigation (Ostrom 2010), and they can help to catalyze broader change.

But what about individuals? Given the magnitude of the climate challenge, individual efforts to reduce their emissions may seem like a drop in the proverbial bucket. Does biking to work or turning down one's home thermostat really make a difference? Perhaps I'm suffering from an outsized view of my own significance if I think my individual actions matter. And if individual actions *don't* make a difference, then one might think that there can't possibly be any individual moral obligations with respect to personal emissions, or perhaps even with respect to climate change more generally. If the problem is large and structural – rooted in the fossil fuel industry, growth-based global economies, ever increasing consumption, and subsidies for oil and gas – then surely the solutions, and the responsibility for implementing them, falls on governments, corporations, and other institutions, rather than on individuals.

In my contribution to this book, I argue that this conclusion is too quick. Responsibility for climate action doesn't fall *only* on governments or corporations, though much of it does lie there.

Individual actions matter morally, and at least *some* people – such as those who care about climate change and have large carbon footprints as well as the capacity to reduce them without significant sacrifice – have good reasons to curtail their emissions. These reasons can be described as *reasons of integrity* and *reasons of relationality*. Reasons of integrity have to do with the connections between one's individual actions and one's values; reasons of relationality have to do with the connections between one's individual actions and one's relationships with others and to the broader community.

Before we get to the details of my arguments, though, I want to offer a bit of background regarding the debate about individual obligations to reduce greenhouse gas emissions. My own engagement in this debate begins at an environmental ethics conference in Colorado, so I'll start there.

2. A Provocation: Encountering an Argument Against Individual Emissions Reductions

More than a decade ago, I attended an environmental ethics conference near Rocky Mountain National Park in Colorado. The conference was wonderful: it wasn't too big, we took hikes in the afternoon, and the presentations were interesting, diverse, and thought provoking. One talk, in particular, really made me think, and one of the reasons it really made me think is because it bothered me. I thought the conclusion was not only wrong, but dangerous. Yet even though I believed that the conclusion *had* to be wrong, I couldn't explain exactly why, so I had to spend more time thinking about the argument, the conclusion, and why I had a different view on the matter.

So, what was this bothersome and thought-provoking argument? It was an argument about climate change, and its conclusion was that *people have no obligations to reduce their personal greenhouse gas emissions in the absence of a collective agreement to do so.* In other words, if a person lives in a society where personal greenhouse emissions are basically a free-for-all (everyone can emit as much as they like, and there are no established laws, rules, or even agreed-upon norms that limit individual emissions), then they have no obligation to reduce their individual emissions.

Climate change is a very serious global problem that is already causing tremendous harm to people and ecosystems globally. Something should be done about it. Given the seriousness of climate change and the imperative of climate action, should individuals really be off the hook with respect to their personal emissions in the absence of a collective agreement to cut greenhouse gases? What's more, if individuals *are* off the hook, won't that potentially lead people to emit *more* than they already do? These were some of my concerns, but before we get to those, let's take a closer look at the details of the argument against individual obligations to reduce climate emissions.

The argument, offered by philosopher Baylor Johnson (2003, 2011), rests on the premise that in the absence of a collective agreement to reduce emissions, individual emissions reductions make little difference. As he puts it, "At least in addressing commons problems, unilateral, voluntary actions typically have no reasonable chance of achieving their object" (Johnson 2003, p. 272). Why might one think this?

Johnson's argument for this premise is twofold. First, he argues, *individual emissions are not harmful*. Because that atmosphere is so large and individual emissions are a drop in the atmospheric bucket, one individual's emissions make no difference: emissions are only harmful in the aggregate (Johnson 2003, p. 273). Second, Johnson claims, *restraint is likely to be exploited*. In the absence of a collective agreement, one individual's restraint may be exploited by others, so individual reductions are very unlikely to reduce emissions overall. Here's what Johnson (2011, p. 154) says about this:

> reductions by one individual may only result in increased use by others. Pretty certainly this is what will happen on an over-crowded common pasture: grass left by the unilaterally reduced herds of A will just be eaten by the herds of B. And if B is not willing to forebear unilaterally, more grass on the commons may inspire her to increase the size of her herd. Something like this seems to have happened in America after the oil embargo of 1973 led temporarily to substantial increases in fuel efficiency, which held down the price of oil and inspired new uses for it, eventually including a huge market for light trucks and SUVs.

Putting the pieces together, Johnson's argument looks like this:

1. An individual has an obligation to reduce their personal green-house gas emissions *only if* those emissions reductions help to ameliorate climate change.
2. In the absence of a collective agreement to reduce emissions, an individual's personal emissions reductions do *not* help to ameliorate climate change.
3. Therefore, in the absence of a collective agreement to reduce emissions, there is no obligation for an individual to reduce their personal emissions.

Johnson (2003) doesn't think that individuals are *entirely* off the hook, however. He believes individuals have an obligation to help establish a collective agreement that would limit emissions. Perhaps this agreement would set rules that constrain individual emissions by limiting the sale of gas-guzzling cars, mandating solar panels on new or renovated houses, or adding an extra tax on fossil-fuel intensive activities and products. With such an agreement in place, people would be obligated to follow its terms – but until then, the obligation of individuals is not to reduce their individual emissions, but to work toward an agreement.

Johnson and I agree that collective agreements and systemic change are needed: changes in rules, policies, and institutions can help reduce emissions and mitigate climate warming, and institutional and structural changes are also needed to help societies adapt (Johnson 2011; Hourdequin 2011). But in the meantime, given the seriousness of global climate change, is it truly no problem, from an ethical point of view, for individuals to drive gas-guzzling SUVs for fun on weekends, as philosopher Walter Sinnott-Armstrong (2005) has argued?

Intuitively, my response to this question was a resounding *no*, but as a philosopher, I knew I needed to figure out and explain why. I also recognized that as I explored the issue, my position might change. But upon reflection, I found some important reasons to question Johnson's argument (Hourdequin 2010), affirming my initial intuitions.

Let's look at some of these reasons, starting with Johnson's first premise:

1. An individual has an obligation to reduce their personal green-house gas emissions *only if* those emissions reductions help to ameliorate climate change.

This premise seems initially plausible. The whole reason we're discussing emissions reductions is due to the climate problem, so if individual emissions reductions don't help to address it, why should anyone be obligated to undertake such reductions?

One answer is this: the first premise seems to presuppose a *consequentialist approach to ethics*, which is not the only reasonable ethical approach. Thus, even if there aren't *consequentialist* reasons to reduce individual emissions, there might be *non*-consequentialist reasons to reduce individual emissions. Let's unpack this further. According to consequentialist ethical theories, consequences are all that matter morally, thus actions or rules can be assessed ethically based on the kinds of consequences they generate. An ethically right action, according to this view, is one that generates good consequences, and an ethically wrong action is one that generates bad consequences.

However, there are sometimes reasons to do things that are unrelated to their consequences. For example, quitting a job with a corrupt firm might be the right thing to do even if quitting won't change the company's dishonest practices. Similarly, protesting a racist policy may be the right thing to do even if one knows that the policy won't be overturned.

Committed consequentialists might, of course, disagree – but the point is that consequentialism is not the only plausible ethical position. For example, protesting a racist policy might be a way of demanding respect for oneself or for a racialized group with which one is identified, or it could be a way to express solidarity with individuals and groups targeted by the policy. Thus, it could be the case that individuals should reduce their emissions, even if those emissions reductions don't directly ameliorate climate change. Driving one's gas guzzling vehicle for fun on a Sunday might be disrespectful to others, or it might be a profligate and unvirtuous thing to do. To provide a full counterargument along these lines would require more detail, but one way to challenge the first premise is to question its consequentialist underpinnings, offering non-consequentialist ethical reasons for individuals to reduce their emissions.

Of course, the first premise could also be flawed from a consequentialist perspective, because there may be consequentialist reasons to reduce personal emissions even if such reductions don't ameliorate climate change. For example, an individual's greenhouse gas emissions might make local pollution worse and aggravate asthma and other health problems for people in the area. These negative consequences provide reasons to reduce personal emissions regardless of their climate impacts.

But since we're focused here on climate change, let's consider further whether there are climate-related reasons to reduce one's personal emissions. Even if the first premise is on target, the second premise might be flawed. Recall the second premise:

2. In the absence of a collective agreement to reduce emissions, an individual's personal emissions reductions do *not* help to ameliorate climate change.

This assumption has generated a massive amount of attention in the literature on climate ethics: Do personal reductions make a difference to climate change, or not? Johnson (2003) says they do not. Sinnott-Armstrong (2005) agrees, using the metaphor of a flooding river: if a person pours a quart of water into a flooding river, they have not caused the flood, and they are not morally responsible for the harm it generates. Similarly, according to Sinnott-Armstrong, if I drive a gas-guzzling SUV for fun on a Sunday afternoon, my emissions are not the cause of climate change, and I'm not responsible for the harm that climate change generates.

Other philosophers disagree. John Nolt (2011), for example, has argued that, *cumulatively*, an individual's emissions can cause significant climate-related harm. By summing up the greenhouse gas emissions of an average American over their lifetime, finding the proportion of climate change to which these emissions correspond, and multiplying this proportion by the total amount of climate-related harm expected to occur in the next millennium, Nolt estimates that the average American "is responsible, through [their] greenhouse gas emissions, for the suffering and/or deaths of one or two future people." (Nolt 2011, p. 3).

Of course, the details matter, and Nolt's calculation is based on a series of estimates, including the estimated emissions of an individual American over their lifetime. Nolt begins with the total annual U.S. emissions, then divides this by the U.S. population to determine average emissions per person. Nolt acknowledges, however, that the *average* per person emissions may not be typical. For example, if a small number of Americans emit quite a lot, and many others emit more modestly, the per person average may be more than most Americans emit. However, in the absence of more fine-grained data, Nolt suggests that the average is a good starting point.

There are other complexities in Nolt's calculation. For example, he estimates U.S. emissions at about 7 billion metric tons. However, in recent years, U.S. emissions have consistently fallen below 7 billion metric tons,[1] and Nolt's admittedly rough estimate doesn't account for dynamic changes in greenhouse gas emissions over time. The overall point stands, however: Americans emit a lot of greenhouse gases, and over the course of our individual lifetimes, those emissions add up.

Nolt's conclusion is provocative – and I believe it is *intended* to provoke by directly taking on the *problem of inconsequentialism* with respect to climate change. Ronald Sandler formulates the problem of inconsequentialism like this:

> [G]iven that a person's contribution, although needed (albeit not necessary), is nearly inconsequential to addressing the problem and may require some cost from the standpoint of the person's own life, why should the person make the effort [to cut their emissions], particularly when it is uncertain (or even unlikely) whether others will do so?
>
> (Sandler 2010, p. 168)

Nolt offers a pointed response to this question, at least for those emitting greenhouse gases at the level of the average American: you should make an effort because *your personal emissions will (indirectly) cause one or two future people to suffer or die.* That sounds bad, and it provides some perspective on the cumulative impacts of a single high-consuming person's emissions over their lifetime. When contextualized globally, it also highlights how much many Americans consume in relation to counterparts elsewhere around the world. As of 2022, Americans emitted, on average, 14.9 metric tons per person (Ritchie, Rosado, and Roser 2023). China averaged 8 metric tons per person, and the world average was 4.7 metric tons (Ritchie, Rosado, and Roser 2023).

In addition, one might go further and look at the contributions of the *wealthiest* Americans, as well as the wealthiest people worldwide, who are responsible for a hugely disproportionate share of global emissions. The emissions of the wealthiest 1% are *not* a mere drop in the global bucket. According to one recent report, in 2019 the world's wealthiest "1% were responsible for 16% of global carbon emissions, which is the same as the emissions of the poorest 66% of humanity (5 billion people)," and in addition, *the wealthiest 10% are responsible for half of all global emissions* (Khalfan et al. 2023, pp. x, 17). Many of these emissions are what Anil Agarwal and Sunita Narain (1991) call luxury emissions: they are emissions associated with activities that may be quite enjoyable, and even edifying – like vacationing in Europe, or flying one's private jet to a TED talk in Vancouver – but they are not necessary to live, or even to flourish as a human being. They are in a different category than the emissions from solid fuel cookstoves (which burn fuels such as wood, coal, or animal dung) on which roughly 2.8 billion people globally rely (Bonjour et al. 2013). The emissions needed to sustain basic life functioning are subsistence emissions, and it seems reasonable that these emissions should be evaluated differently than luxury ones (Shue 1993). Cutting subsistence emissions undermines people's basic life functions, survival and flourishing, whereas cutting luxury emissions might require some lifestyle changes, but fundamental needs are not at stake. Although Nolt doesn't make the calculation, the personal emissions of the highest consumers globally are significantly greater than those of the average American, with an even bigger climate impact.

With all this in mind, it seems that – at least for those of us who are relatively wealthy with consumption to match – individual emissions are not inconsequential. Of course, as Nolt himself acknowledges, it is not possible for me to trace the effects of my emissions to any *particular* harm to any particular person in any particular place, but that doesn't mean that my contributions are not damaging. Nolt's answer to the problem of inconsequentialism is to say that all of us – and especially those who emit the most – are responsible for a share of the climate problem and a share of the harm it causes. And although institutions and infrastructure constrain the degree to which it is possible or easy to reduce one's personal emissions (if I live in an area with poor public transportation and can't afford to live close to where I work, it may be difficult to avoid driving a car to my job), many people – especially wealthy ones – have substantial control over their luxury emissions. Especially once subsistence needs are met, many of us can modulate our contributions to climate change, and those who are relatively privileged often have choices when it comes to subsistence emissions as well. These include options regarding how the size of one's living space, the kinds of foods one chooses to buy, and so on. Over time, individual choices add up.

Even so, some will object that Nolt's analysis is abstract and theoretical. Can we really take a huge problem like climate change, caused by the cumulative effects of emissions in so many different places and by so many different people and institutions, and simply determine causal and moral responsibility by the kind of aggregation and division Nolt employs? Some argue that responsibility for climate harms can't be apportioned in this way. As Jason Kawall (2011, p. 23) asks, "[I]s responsibility for one two-billionth of the suffering or death of two billion people always morally equivalent to responsibility for the suffering or death of one person?"

Certainly, the causal relationships between an individual's emissions and the harms that result from increased greenhouse gases in the atmosphere is more complex and indirect than in the case of "paradigmatic moral harms," where one identifiable individual directly harms another (Jamieson 2010). There are complex philosophical questions here involving causation and moral responsibility, but there is an important lesson from Nolt's analysis: although climate change is the result of the combined effects of many people's and institutions' actions over time, neither the problem nor the solution is a binary, all-or-nothing one. Incremental contributions matter. Recent reports produced by the Intergovernmental Panel on Climate Change (IPCC) – the scientific organization that synthesizes climate science globally – underscore this point. Climate impacts will be significantly worse at 1.5°C of warming than they are now, and they'll be even worse at 2°C (IPCC 2018a, p. 5). Every ton of carbon added to the atmosphere matters (cf. IPCC 2018b, p. vi).

With all that said, I don't think that the only reason to curtail one's personal emissions is due to the kind of incremental harm-reduction that an individual's emissions reductions might achieve. Relying too strongly on this approach has the potential to lead us into the weeds of arguments over whether one person's restraint will be compensated by another's greater use (a possible, but not inevitable, outcome), and over the precise consequences of a particular increment of emissions. Notice, for example, that Nolt (2011, p. 9) tells us that the average American's emissions will cause "the serious suffering and/or deaths of two future people," but that description marks out a pretty wide range of outcomes: is it suffering, or death? How much suffering? If my lifetime emissions cause two people over the course of the next millennium to suffer, that's one thing, but if my lifetime emissions cause two people to *die*, that seems like a different and *much* more serious outcome. If I knew that cutting my luxury emissions by 50% would save a person from *dying*, I'd like to think I'd work very hard to reduce those emissions. On the other hand, if those same reductions would merely alleviate the *suffering* of two people (Nolt says "serious suffering," but this is very vague, and it's unclear whether this predicted suffering is time limited or lifelong), the ethical pull of personal emissions reductions feels weaker and more diffuse, especially since *much* of what I do causes some suffering, somewhere. Even if I eat a vegetarian diet, I depend on agricultural crops, and the plows and combines moving through farm fields kill small mammals like mice (Cahoone 2009, pp. 80–81, citing Davis 2003). And even if I try to buy secondhand clothes, somewhere along the line, production of the shirt I'm wearing likely involved the labor of someone working in very poor conditions. In this sense, most of us are complicit in causing some amount of suffering in the world (for a range of perspectives on complicity and its moral implications, see Kutz 2000; Young 2006 – who does not use the word "complicity," though her work clearly bears on it; Lawson 2013; MacLean 2019; Galvin and Harris 2023). I want to reduce this suffering where I can, but the particular increment of suffering I cause through my personal greenhouse gas emissions is hard to connect to actual impacts on the ground, which can make the whole idea seem a bit fuzzy.

This fuzziness may help *explain* why it can be hard to motivate action on climate change, but it doesn't necessarily *justify* inaction. And the diffuseness of impacts associated with individual (or even institutional) choices can open the door to what philosopher Stephen Gardiner (2006) calls moral corruption, a set of strategies for rationalizing and justifying (relative) inaction on climate change. Gardiner sees this operating at the international level, in the development of weak agreements with few accountability mechanisms that are nevertheless celebrated as great achievements. Gardiner (2022) argues that the Paris Agreement is one example of this at the international scale. In the individual realm, the fuzziness may make it easy for a person to make some very modest lifestyle adjustments (I recycle and compost!), then think their work is done.

In addition, the fuzzy connections between individual actions and their climate impacts can make it hard to see climate change as a moral problem in the first place (Jamieson 2010). According to Dale Jamieson (2010), climate change doesn't have the structure of what he calls a "paradigmatic moral problem," A paradigmatic moral problem is one in which one person does something that harms or hurts another, and the first person's actions are clearly and directly connected to the second person's harm or hurt. Jamieson gives the example of one person stealing another person's bicycle. The cause of the harm is the theft, and the person harming and the person harmed are both clearly identifiable. But my greenhouse gas-emitting actions are not like this. Many of them – considered individually – contribute negligibly to climate change, as we've discussed. Although individual actions add up over time, their impacts on the climate system can't be easily identified: Was it *my* lifetime of wasteful consumption that pushed a specific Gulf Coast hurricane from category 4 to category 5? Over the last decade, there have been significant advances in attribution science, which can link extreme weather events to climate change, assessing the degree to which a certain amount of climate warming makes a particular event more likely or more severe. However, even the most sophisticated attribution science can't link a *particular* person's carbon dioxide emissions or a particular release of methane to a specific weather event. And without this kind of linkage, it's hard to recognize one's personal carbon emissions as fitting the example of a paradigmatic moral harm.

And yet, even though climate change may lack certain paradigmatic features, many people today clearly recognize it as a moral problem. Climate impacts are significant and highly visible globally, and the causes – even if widely distributed – are well known. Where things get trickier is in the attribution, both causally and morally. Which actors, specifically, should bear responsibility for climate change, and what kinds of responsibilities should they bear? My view is this: although it's important to recognize that individual emissions contribute to climate change, especially through habits and lifestyles maintained over time, it doesn't make sense to spend too much energy partitioning climate responsibility based on each individual's particular quantity of carbon emissions. Not only would this involve lots of complicated calculations; from an ethical perspective it would also require acknowledgement of and distinction between subsistence and luxury emissions, and one might also want to take into account each person's particular life circumstances and constraints, which can make it more or less difficult to curtail emissions (Baatz 2014). Moreover, since the linkage between one's own emissions and particular, specific harms is hard to draw, it may be hard to morally motivate people by trying to make these links.

Approaches along the lines of Nolt's may help make the connections more vivid, and carbon footprint analyses (see, e.g., www3. epa.gov/carbon-footprint-calculator) can help people identify places where they can reduce their emissions, but there are additional reasons to consider reducing one's emissions that don't focus solely on the direct climate impacts of those emissions. One reason people who are concerned about climate change may think that individual emissions don't matter is because they focus too much on the *direct* effects of those emissions, which are hard to trace, and too little on the ways in which reducing emissions may be connected to broader personal values, to a different way of conceiving the relationships between individuals and collectives, and to deeper social transformations.

These latter ways of thinking about individuals' roles in responding to climate change do not fixate on or fetishize individual reductions, but instead see individual thoughtfulness about greenhouse gas emissions as part of broader social and cultural shifts toward sustainability, mutual care, and collective flourishing. Additionally, they focus less on backward-looking responsibility (who did what, who should bear responsibility for the consequences – though such approaches remain important in climate contexts, especially at the level of corporations and nations) and more on forward-looking potential for change (for discussion of forward and backward responsibility, see Young 2011). They link taking responsibility to figuring out how one is empowered to work toward responses that bring people together to effect change rather than isolating them through individualized guilt and responsibility (again, see Young 2011, for related discussion). Taking individual responsibility doesn't have to distract us from holding corporations or governments accountable; instead, it can support accountability at multiple levels.

Recall the second premise of the main argument against personal emissions reductions:

2. In the absence of a collective agreement to reduce emissions, an individual's personal emissions reductions do *not* help to ameliorate climate change.

We're now in a position to question this premise by pointing out that it can be interpreted in two different ways, both of which can be challenged. The first interpretation focuses on the direct impacts of personal emissions reductions, holding that reducing one's personal emissions does not directly ameliorate climate change. As we've seen, whether an individual's emissions directly exacerbate climate change is controversial. The inconsequentialist view suggests that individual emissions don't matter in the grand scheme of things. I have pushed back against this position, drawing on Nolt's analysis to suggest that individual emissions *do* matter from a climatic perspective.

However, I have also acknowledged that the specific impacts of individual emissions are fuzzy, and we'll likely never know how our own emissions mix with those of others to impact the climate system or understand the precise downstream effects on people and ecosystems. Thus, it can be hard to see how driving a car more or less, or turning up the heat on a cold winter day, contributes to specific climate-related harms. Given this fuzziness, people may be prone to "explain away" their emissions – even those that clearly fall into the category of luxury emissions – as necessary or impossible to avoid. Focusing on how individual emissions exacerbate climate change may therefore activate various forms of moral corruption, such self-deception and complacency (for discussion of moral corruption and climate change, see Gardiner 2006, 2011). In addition, focusing too much on individual contributions to climate-related harms may activate *guilt* as a primary moral motivator. Although guilt can promote pro-environmental behavior under certain circumstances (see, e.g., Rees, Klug, and Bamburg 2015; Hurst and Sintov 2022), it may also trigger responsibility-avoiding defensiveness and "reactance" (Graton, Ric, and Gonzalez 2016) rather than responsibility-taking climate action. Forms of eco-guilt and responses to it are complex.

With these caveats and complications in mind, we can see some of the limitations of focusing *only* on the direct effects of personal emissions, even it is true that an individual's personal emissions reductions *can* directly ameliorate climate change (despite the difficulty in tracing the extent or specific benefits of an individual's averted emissions). Therefore, I think another interpretation of the main argument's second premise deserves a closer look.

A second interpretation of the premise is broader and stronger than the first: it holds that in the absence of a collective agreement to reduce emissions, an individual's personal emissions reductions do not help to ameliorate climate change *in any way*. Even if Johnson were right that individual emissions reductions make no direct difference to the climate, the second version of his premise might be false: it might be the case that individual emissions reductions make a difference not so much through powerful ameliorative impacts on the climate system, but through their positive impacts on *social* systems that are important in mediating broader responses to climate change (Hourdequin 2010). As I explain below, I think individual emissions reductions can have exactly this sort of positive effect. If that's right, then we should consider not only how individual emissions affect the climate system directly, but also how people's climate-related actions enable or hinder constructive responses to climate change at multiple levels. To be clear, at no point in this book do I argue that the sole or main focus of individual climate action should be on individual emissions reductions. However, I do think that there are good reasons to consider one's own emissions as part of one's broader thinking about individual ethical responsibilities in relation to climate change, as I go on to discuss.

Summing up what we've considered so far, philosophers have offered some important and influential arguments in support of the conclusion that individuals lack obligations to reduce their personal emissions in the absence of laws, rules, or a collective agreement to do so. One key line of argument suggests that individual emissions have no meaningful climate impact, and that without such impact, there's no reason for individuals to make emissions cuts. However, individual emissions *can* have direct impacts on climate, and they can also have *indirect* impacts on responses to climate change. Moreover, there may be reasons to reduce one's personal emissions that aren't directly tied to the specific consequences of doing so.

In the sections that follow, I offer two key kinds of reasons for individuals to attend to their personal emissions, though – as I will explain – I believe that individual action in relation to climate change should not focus only or even primarily on personal emissions reductions, and what you, or I, or others should do about climate change depends importantly on our individual contexts.

3. Why Do Personal Emissions Matter? Reasons of Integrity

3.1. From Hypocrisy to Integrity

Individual emissions are a fraught topic in climate debates, within and beyond academia. For many years, environmental movements in the U.S. and elsewhere focused on individual action, guided by slogans like "think globally, act locally" and "reduce, reuse, recycle." Exhortations to individual actions such as buying local organic produce, swapping out incandescent bulbs for compact fluorescents and (more recently) LEDs, going vegetarian, and using recycled paper have been the stock and trade of environmentalism for decades. In 2009, author Colin Beavan documented his family's efforts to live lightly on the planet in the book and later film, *No Impact Man*; a 2007 book – *Plenty* – by Alisa Smith and J. B. McKinnon lauded the environmental benefits of eating locally through a "100 mile diet"; and more recently, youth climate activist Greta Thunberg made headlines for sailing across the Atlantic to the UN Climate Conference in a "zero-carbon" boat (Watts 2019).

But not everyone thinks this is where environmental or climate-conscious action should be focused, and there's been significant backlash against the idea that individuals ought to minimize their environmental footprints. Those who tout the importance of individual emissions reductions, or who make significant efforts to limit their own emissions, are often accused of naiveté, sanctimonious virtue signaling, or an excessive focus on moral purity (for discussion, see Heglar 2019; Zaraska 2022). Climate activists "caught" driving to the store or traveling by air are accused of hypocrisy: if they're so committed to ending reliance on fossil fuels, why are they using them? The arrows sling from multiple directions. From the left, those focused on the structural responses to climate change worry that carbon footprint analyses and corporate campaigns to popularize them deflect attention from the real sources of the problem: energy companies and other industries whose profits are linked to fossil fuels. From this perspective, those of us who dutifully click through the "Know Your Carbon Footprint" website sponsored by BP are simply dupes, falling for corporate propaganda that aims to make *individuals* feel guilty for problems that *corporations* cause. On the other side, some conservative media mock those who bring their own bags to the grocery store, recycle, or forswear long-haul flights, dismissing these efforts as mere virtue signaling: engaging in conspicuous behaviors and pronouncements aimed to demonstrate their moral rectitude. The accusation of virtue signaling comes from the left, too, from those concerned that a focus on individual "green" living may lead to political complacency (Heglar 2019). The worry is that people may think that as long as they reduce their personal environmental impacts and have achieved some modicum of environmental purity, they've done their part. Also from the left comes the argument that exhortations to reduce one's own environmental impact may reflect inattention to people's diverse social locations in relation to climate change. Is it fair or reasonable to expect people of limited financial means to buy new LED lightbulbs or invest in solar power for their homes? Maybe making green choices – buying electric cars, locally grown produce, and the like – is just another way for wealthy people to flaunt their privilege (Monbiot 2007).

These debates reveal that individual emissions choices are themselves entangled with the complex politics of environmentalism and climate change Those who criticize a focus on individual emissions reductions don't come from one part of the political spectrum, and the contestation raises the question of what's really at stake behind all the debate. The entanglement also suggests that individual emission choices reverberate beyond their effects on the climate system: they have political meaning, and this meaning is interpreted differently by different groups.

From an ethical perspective, how might we think about these debates? Taking account of the criticisms of a single-minded focus on individual emissions reductions, what reasons might there be for individuals to consider their own contributions to climate change?

My view is that individual emissions matter morally for (at least) two kinds of reasons – reasons of integrity and reasons of relationality – and (all else equal) these reasons count in favor of individual emissions reductions. Reasons of integrity apply most directly to those people who already care about climate change and who are committed to doing *something* about it, whereas reasons of relationality apply more generally. Ultimately, I will argue that reasons of integrity are relevant to people with a very broad range of values and concerns – and not just those who have special concerns about climate change – but that argument comes later in the book.

We can get an initial, intuitive handle on reasons of integrity by beginning with hypocrisy. Hypocrisy comes from the Greek word *hupokrisis*, referring to acting or playing a theatrical part. This etymology suggests that hypocrisy may involve a kind of pretending or pretense. For example, a person who regularly professes their commitment to honesty yet lies at every turn might merely be *pretending* to care about honesty. But hypocrisy isn't only about pretense. It can also involve calling out others for faults one shares (captured in the idioms "that's the pot calling the kettle black" or "people who live in glass houses shouldn't throw stones"), which can be characterized as a kind of problematic or wrongful blaming (Crisp and Cowton 1994). Or it can involve inconsistency between words and deeds, between one's professed moral commitments and the way one lives their life (Crisp and Cowton 1994). Lastly, hypocrisy may involve a kind of complacency, where a person "[takes] morality seriously in very unimportant ways, ignoring its demands where their fulfilment appears costly" (Crisp and Cowton 1994, p. 345).

Arguably, all these forms of hypocrisy are bound up with climate change, but I want to focus on how concerns about hypocrisy play out in relation to individual values and actions. I am going to set aside concerns about hypocrisy that are used disingenuously to discredit climate activists (e.g., accusations that suggest that *any* dependency on fossil fuels is hypocritical for those who care about the climate), and instead focus attention on what seem like more genuine concerns. Take the case of a celebrity climate activist who touts the importance of decarbonization, but flies hundreds of times a year in their private jet, typically for brief jaunts to their second or third home. The vast majority of these private jet trips unquestionably involve luxury emissions, and the celebrity (at least in our imagined example) could easily fly less. One might identify two sources of hypocrisy here: (1) the celebrity's actions don't align very well with their professed values, and (2) they don't seem to be taking their professed values particularly seriously. In Crisp and Cowan's terms, their lifestyle seems to reflect a kind of moral complacency about climate change, *even if* they publicly support climate action. A celebrity in such circumstances might argue that they need to fly frequently to participate in important gala fundraisers, maintain their visibility at public events, and so on, but these responses ring a bit hollow if the person in question is not making *any* effort to reconcile their climate-related concerns with their profligate greenhouse gas emissions, especially because our putative celebrity clearly has ample resources and power: there is nothing obviously constraining their ability to live in greater alignment with their values. What's more, since celebrities and other individuals with high socioeconomic status influence others' behavior (Nielsen et al. 2021), the visible and excessive use of private jets conveys the idea that flying frequently is desirable. In the case of our imagined celebrity, their private jet use may further suggest that climate change can be addressed without any significant change in highly consumptive lifestyles.

With all of this in mind, it seems like the celebrity climate activist has reasons to attend more closely to their personal emissions, and to consider what taking more seriously their professed values might involve, in terms of personal action. It may already be the case that the celebrity in question is raising awareness about climate change, supporting policies that would accelerate emissions reductions, and stumping for candidates for whom climate adaptation and just transitions away from fossil fuels are a priority. My point is that they may still have reason to consider and reduce their own profligate emissions. It is not as if taking responsibility in one domain cancels out responsibilities in others: if I'm kind and respectful to my students, that is not license to be rude and dismissive to my coworkers, and if I care about equity, I shouldn't only advocate for more equitable laws; I should also develop equitable practices in my daily life and work.

Although the idea of hypocrisy can be illustrative, I don't believe that the best way to justify or motivate individual climate action – whether at the political or more personal level – is to exhort people to avoid hypocrisy. "Don't be a hypocrite!" is an admonishment that generally comes after something has already gone wrong. Focusing on hypocrisy tends to emphasize the negative, highlighting reasons that count *against* certain ways of acting or being rather than reasons that count in their favor. Hypocrisy's flip side, at least in many respects, is integrity, and I want to argue that there are reasons of integrity to take seriously one's individual greenhouse gas emissions – though these are also reasons to take seriously one's climate-related actions *generally*, so reasons of integrity don't suggest that individual emissions reductions as the only actions that matter, or that reducing their carbon footprint is the most important action an individual can take.

Integrity can be a bit of a slippery concept, used in different ways in different contexts, so I'll soon say more about the specific conception I have in mind. But first, I want to give a basic overview of why integrity is relevant to individual climate action.

Integrity, on my view, is a virtue that has to do with the relationship between values and actions, as well as the relationships among diverse values. It presses us to live by our values, and to harmonize – to the extent possible – the values we have (Hourdequin 2010, pp. 447–451). It does not require fetishizing a kind of fastidious consistency, but it does involve taking seriously one's moral commitments and trying to live in a way that embodies them. With this in mind, I want to suggest that those who are concerned about the social, ecological, and planetary impacts of climate change have reason to live and act in alignment with this concern. Doing so can involve actions at multiple levels.

3.2. What Is Integrity, and Why Does it Matter?

The core idea of integrity has to do with wholeness (Cottingham 2010; Hourdequin 2010). To integrate different ideas is to bring them together into a coherent whole, and to integrate one's values and commitments involves something similar. Integrity also carries connotations of solidity and reliability: a person of integrity is someone we can rely on to honor their promises and to live by their values.

Philosophers Robert Audi and Patrick Murphy (2006) suggest that there are two aspects to integrity: one aspect involves bringing together multiple values and commitments, or multiple aspects of one's character (this is what Audi and Murphy call *integration*) and the other involves internalizing one's commitments and making them deeply a part of – or *integral* to – who one is and seeks to be. Integration, they explain, helps a person avoid conflicts among their various commitments; it involves "a kind of unity among the elements in which they form a coherent, ideally a harmonious, structure" (Audi and Murphy 2006, p. 9). The second aspect of integrity makes one's values and commitments integral, such that they guide one's thought and action, enabling a person to act consistently on their values, often without significant deliberation (Audi and Murphy 2006, p. 9). A person of integrity, on this view, has a set of values that not only fit together into a coherent whole, but that also find expression throughout that person's life.

Why care about integrity? What makes it a virtue, or a trait of character worth developing? Why might living or acting with integrity be a good thing to do? As a virtue, integrity has both personal and interpersonal value (Hourdequin 2010). It has personal value in that it draws together conviction and action, and it encourages individuals to find ways to live that enable them to harmonize, honor, and act in alignment with their deepest values. It also includes reflection on those values, and revision and refinement of values and commitments over time, in light of new information, insight, or understanding. Integrity has the potential to enrich a person's life by reducing internal conflict and helping them organize their life in a way that meaningfully expresses their core values and supports that which they care about. Integrity can be an antidote to fragmentation and dis-integration (Cottingham 2010), and it can be grounding, linked to a sense of purpose and direction.

Integrity also has interpersonal dimensions. For example, engagement with others can help support individual integrity by enabling us to reflect on and readjust our values in light of theirs. Responsiveness to others can enable development of individual integrity, because integrity does not seek shallow consistency, but a deeper harmonization of values reflectively held. The interpersonal value of integrity ties in part to intelligibility to others (Hourdequin 2010, p. 451), and to reliability. A person who lacks integrity has neither made sense of their values nor aligned their values with their actions. They may be honest in one context and dishonest in another, simply because a commitment to honesty is not integral to who they are. Without a clear sense of how their values fit together, they may not know why they do what they do, and thus they may be susceptible to manipulation for others' nefarious ends. Interpersonally, integrity supports a kind of consistency that makes us intelligible to one another, and integrity helps affirm to others the authenticity of one's commitments. Through its links to reliability and genuine commitment, integrity may foster interpersonal trust, which can be important to relationships of mutual support and solidarity.

Integrity thus matters because it can support individual flourishing and enable us to live meaningful lives that align with our values, and because it is important to interpersonal relations, including relations of trust and solidarity.

3.3. Integrity-Based Reasons for Personal Emissions Reductions

Integrity, as I have been describing it, is a virtue, a positive character trait: people with integrity express this virtue in their thoughts, actions, and ways of relating to others. A person who is committed to climate action *and* who fully possesses the virtue of integrity is someone who has *integrated* their climate commitments with their other values to the extent that they can, and who takes their climate commitments (and other vales) as *integral* to their life: their values shape not only what they say, but what they do and how they live their life. Many of us know people who have strongly held values that they have integrated with one another and made integral to their lives. If we are fortunate, we have encountered teachers or mentors, for example, who genuinely care for their students or mentees, and whose time and energies reflect this care. Many who work in health care remain committed to their patients and reflect this commitment in their work despite structural obstacles that push them to hurry through appointments and leave little time to write detailed notes afterwards. Integrity in relation to environmental values, such as a concern to minimize the harm of climate change, can be reflected similarly: some climate scientists, for example, are not only writing books and raising public awareness about climate change, but also pledging to reduce their personal air travel – including travel to academic conferences – and lobbying for changes in the profession that enable colleagues to connect with one another through less carbon-intensive means (see, e.g., https:// noflyclimatesci.org).

Of course, for many of us – or perhaps most – developing virtues such as integrity remains a work in progress. However, even if we don't fully possess integrity, integrity can provide reasons that support certain kinds of choices and actions. Integrity, like other virtues, is associated with certain kinds of reasons for action, and people who possess a particular virtue are "[disposed] to respond to reasons characteristic of that virtue" (Saunders 2021). For example, as Saunders (2021, p. 2786) explains, "to say that someone acted out of generosity, then they must have acted on some reason characteristic of generosity, such as 'they needed the help,' or 'they are my friend.'" In relation to climate change, I have suggested that those committed to climate action have *reasons of integrity* to consider their personal emissions. But what does this mean? How might reasons for action be related to a virtue such as integrity?

Reasons characteristic of integrity follow from the two dimensions of integrity – intrapersonal and interpersonal – discussed earlier. The fact that a particular action would help to harmonize one's values, or align one's values and actions is – with respect to integrity – a reason to act. Similarly, the fact that an action would show a kind of consistency of commitment that would make my actions intelligible to others, or that would help to promote relations of trust, also constitutes a reason for action from the perspective of integrity.

For those of us who genuinely care about climate change, reasons of integrity support a shift from a kind of complacent and passive concern about climate change to an active commitment to do something about it. Some scholars have characterized this passive concern – concern that is not really integrated into one's life and actions – as a kind of climate denial: not a blatant denial that climate change is happening or that it is human caused, but an unwillingness to truly face and respond to the problem (Chislenko 2022; see also Norgaard 2011). In a short story about climate change, "A Full Life," Paolo Bacigalupi (2019) describes this denial as a kind of pervasive "pretending." In the story, even those aware of the increasing seriousness of climate change seemed to hope that "if they pretended really, really hard, they'd be okay" (Bacigalupi 2019). For those of us who profess to care, integrity offers reasons to move from climate denial to climate action. Integrity does not *only* provide reasons for personal emissions reductions by the climate concerns: instead, it supports all kinds of actions to address climate change, and it provokes us to consider how we are best positioned to help. Integrating concern about climate change with other values requires thinking about how to care for oneself and one's friends and family while *also* considering how to care for climate-affected communities and ecosystems around the world. It requires thinking about how to transform economies, institutions, and cultures in a way that enables current and future generations to flourish. And it requires thinking about what each of us can do, at multiple levels, to effect constructive change.

Interestingly, although Baylor Johnson (2003) and Walter Sinnott-Armstrong (2005) have both argued that individuals lack obligations to reduce their personal emissions in the current context, they nevertheless believe that people have *some* climate-related obligations. These obligations, they each suggest, focus primarily on work at the political level to establish laws and policies that limit climate change, and to elect leaders who will enact such laws and policies. If they are right, then individual obligations are political obligations *only*. My view is that – in general – if people have political obligations concerning climate change, then they *also* have obligations to consider their personal emissions (even if those obligations are defeasible – that is, even if they can be overridden by other considerations). These obligations, as I have suggested, are grounded in reasons of integrity. However, the idea that reducing greenhouse gas emissions is the government's job – and not the responsibility of individuals – is widespread, so it is important to address it explicitly.

3.4.   Is Climate Change the Government's Responsibility (Only)?

For Sinnott-Armstrong (2005), the greatest responsibility for climate action falls on governments, particularly national governments like that of the U.S. The United States is a wealthy country with significant capability to do something about climate change, and U.S. emissions – both historically and continuing today – are a major cause of climate change. The U.S. government has the wealth, the power, and technological capacity to mitigate and facilitate adaptation to climate change: the government is *em*powered to make a difference. However, according to Sinnott-Armstrong, just because the U.S. government should act to reduce emissions nationally doesn't mean that *each individual* in the U.S. has a corresponding obligation. As he puts it, "individual moral obligations do not always follow directly from collective moral obligations." (Sinnott-Armstrong 2005, p. 295). A key aspect of Sinnott-Armstrong's view is that national obligations to cut emissions can't be neatly partitioned into individual obligations to do so: "The fact that your government morally ought to do something does not prove that you ought to do it, even if your government fails" (Sinnott-Armstrong 2005, p. 295). Sinnott-Armstrong gives the example of a damaged bridge, and one might broaden the example to consider decaying infrastructure more generally. If there are potholes in the street or a damaged bridge that is unsafe to cross, it's the government's responsibility to fix them. If the government doesn't fix them, people should hold the government accountable, but they don't have a moral obligation to fix the potholes or repair the bridge themselves – or so Sinnott-Armstrong argues.

At first glance, this seems like a very reasonable position. After all, it is the government's job to fix roads and bridges, and to ensure that the basic infrastructure of society remains intact. Similarly, argues Sinnott-Armstrong, it's the government's job to address climate change. However, climate change is disanalogous to a decaying bridge in multiple important ways. First, both the causes of climate change and the responses that can help alleviate it are more diffuse and distributed than in the bridge case. While it is true that the causes of the bridge's deterioration are multiple and distributed across time, they are not nearly as multifarious as the causes of climate change, and with respect to repair, the rebuilding of the bridge is a much more bounded and tractable project for a government or an arm of government to address than preventing dangerous climate change. In this sense, it is more appropriate to charge governments with the primary or sole responsibility for fixing bridges than it is to expect governments alone to fix climate change. Second, keeping bridges in good repair doesn't seem to require a fundamental reorientation of basic institutions, values, and norms in many current societies, whereas adequately addressing climate change might (for one such argument, see Klein 2015). Such reorientations generally require change at multiple levels, including cultural change. Finally, many governments – at least in reasonably well-functioning municipalities, states, and nations – have a successful track record in maintaining bridges and other infrastructure, but a much weaker record in responding to climate change. Thus, although I agree with Sinnott-Armstrong that people *should* pressure governments to do more about climate change, it's not clear that governments alone have the capacity to adequately address it. Responding adequately to climate change requires multiple actions by multiple actors at multiple scales. Individuals who are concerned about climate change should thus consider how their actions can contribute to a multi-scalar and polycentric responses to climate change (Ostrom 2010; Vanderheiden 2016), and they should also consider what acting consistently with their values might require in multiple domains, including their personal lives.

As I argue below, interpersonal relations are not so distinct from larger fabric of culture and institutions: there is no sharp line between the personal and the political, or between individual and collective actions. People are connected to one another and to the broader world through complex networks of relations, and ethical responsibilities are embedded in these relations. Thus, as I argue below, from a relational perspective, personal emissions matter.

4. Why Do Personal Emissions Matter?
Reasons of Relationality

In some parts of the western United States, such as Las Vegas, there are water waste investigators who patrol the streets looking for leaky sprinkler heads or rogue homeowners who try to sneak in an extra watering session to keep their lawns lush and green (Fernandez 2022; Marszal 2023). In these areas, water is such a scarce resource that cities and towns have established strict rules to curtail household consumption and make sure there's enough to go around. The rules don't generally restrict the use of water for cooking, bathing, or general household use, but they do apply to yards and lawns.

These rules – even if they're not universally endorsed – might be seen as the kind of collective agreement Baylor Johnson has in mind in relation to climate change. Following his logic, once such rules are in place, people have an ethical obligation to follow them, but in the absence of such rules, they don't. According to this view, individual water conservation – even in water-limited areas further taxed by severe drought – is optional in the absence of a societal agreement to curb water use.

Although water shortages in the American West aren't exactly like climate change (though they are linked to and exacerbated by it), there are some key similarities in relation to individual action. Here in Colorado Springs, if I water my lawn less or swap out the bluegrass for drought-tolerant plants, the impacts on the Colorado River watershed as a whole are minimal. Will my restraint help save fish downstream? Will it prevent a farm in Arizona from going belly up? Will it enable the river to make it just a bit closer to the Gulf of California before running dry? It's hard to know. But if I transform my bluegrass lawn into a beautiful garden of plants that thrive in dry conditions, maybe my neighbors will ask about my garden, and maybe when I tell them that it's low maintenance and was easy to plant, they'll look into exchanging *their* lawn for a xeriscaped yard.

Of course, it's *possible* my water conservation will just leave more for others to use (as Johnson suggests is the case with respect to personal greenhouse gas emissions reductions), but it doesn't seem the most likely outcome. Social dynamics are complex, and people aren't driven only by the desire to consume more of everything. Non-functional grass lawns in the arid West are – in a sense – the equivalent of luxury greenhouse gas emissions: they may be nice to have, but people can live perfectly well without them. As with greenhouse gas emissions, water use is not a binary thing: some use is critical to people's lives and flourishing, but some of it is not. Water conservation doesn't necessarily make people worse off, and in some cases, it may make people's lives better.

The ubiquity of the American lawn is based in certain aesthetic values that emerged in Europe and migrated with settlers across the Atlantic. However, there are other approaches to landscaping that can be aesthetically pleasing. Water consumption for lawns is, in large part, a cultural thing. Americans are "lawn people," as author Paul Robbins (2007) suggests. What's more, the cultural appreciation of lawns is bound up with the *politics* of lawns: even in the absence of *rules* about how much or how little to water one's lawn, there may be all sorts of *expectations* in place. The lack of a city ordinance against dandelions doesn't prevent neighbors from looking askance at the weedy yard on their suburban block, and it doesn't prevent people from feeling ashamed of their own weeds or compelled to assault them with herbicides to maintain the mono-cultural peace.

However, there are efforts underway to disrupt the dominant culture of the lawn in the United States. Here in Colorado, for example, the nonprofit organization Resource Central offers a "Garden In A Box" program that enables homeowners to purchase a waterwise garden kit designed by landscape professionals (https://resourcecentral.org/gardens). These kits come with a set of plants and a "plant by number" guide, showing where to locate plants to fit the garden plan. The plants are perennials, and once established, they thrive year after year. These attractive gardens have the poten-tial to help develop an appreciation for landscaping that diverges from the standard green grass lawn, and they can thrive with min-imal supplementary water, even during periods of hot, dry weather. Low-water gardens can also save time and money: they don't need to be mown, and lower water use means lower water bills.

But what does all of this have to do with climate change? Low-water gardens are one way of adapting to increasingly warm and dry conditions in some regions, and individual choices to install these gardens can be a catalyst for broader social change to sup-port water conservation and adequate water resources for all. Landscaping one home with low-water plants won't resolve water scarcity issues in the U.S. West, but if each garden-in-a-box planted inspires a couple of others to do the same, the effects can multiply. This illustrates how individual actions may be beneficial not only through their *direct* effects on the problem at hand, but through the *indirect* effects of individual actions on others' decisions, as well as broader social values and norms. There are ethical reasons for people to plant a waterwise Garden In A Box that go beyond the amount of water saved in their individual yards.

What's more, not only may such individual actions cata-lyze others to make similar choices; they may also set in motion broader social change.[2] Neighborhoods previously full of uniform green lawns may be transformed to more diverse landscapes, shifting the expectations around appropriate landscaping. People who live in adjacent neighborhoods where lawns are required by their homeowners' associations (HOAs) (and there are many such neighborhoods, especially in wealthier suburban areas) may enter into conversations with their HOAs about revising these rules. If demand for waterwise gardens increases, nonprofit organizations like Resource Central may expand their programs to new areas or develop partnerships with local water providers to provide rebates or discounts on garden purchases. This, in turn, may lead more homeowners to shift to xeriscaping, creating a virtuous cycle. These are not pie-in-the-sky ideas: these kinds of shifts are already happening in Colorado and throughout the Western U.S.

Similar processes might be sparked by actions that reduce individual emissions. A couple of people carpooling, or a single person walking or biking to work rather than driving, won't change the course of the global climate system. But maybe many of those who shift to carpooling will find that it's more pleasant to have company on the drive to work rather than driving alone, or maybe carpooling can make parents' busy lives easier in getting kids to and from school. As more kids in a neighborhood walk to school, walking becomes a safer option and parents may worry less that their kids are the only ones walking.

These kinds of expanding social impacts offer support for individual emissions-reducing choices, and they are impacts that should matter to ethical consequentialists. From a consequentialist perspective, *all of the consequences* of an action matter. Thus, if it's true that individual emissions reductions can catalyze further reductions as well as climate-protective social change, individual emissions reductions are ethically important. Following this line of argument, in the absence of countervailing negative consequences, individuals – insofar as they can – should reduce their emissions. With caveats about subsistence emissions and different contexts in mind, we might conclude that all else equal, it is ethically better for individuals to emit less rather than more.

As we will see later in this book, this argument encounters potential objections when one goes all-in on consequentialism. If one considers *all of the possible actions* a person might take to make the world a better place, it becomes harder to argue that personal emissions reductions are *the most important way* to contribute. However, my intention is not to argue that personal emissions reductions are the only or even the best thing one can do to make the world a better place, nor do I think that consequentialism is the only or best way to think about the ethics of climate change. My view is that *there are good ethical reasons for people to consider reducing their personal emissions as part of broader consideration of how to contribute constructively to climate action.* I have already discussed how reasons of integrity may support both individual emissions reductions and broader social and political action on climate change; here, I want to discuss *reasons of relationality,* and how the observations above – which suggest that individual actions are embedded in broader social contexts and can catalyze broader social change – support not just consequentialist reasons for climate action, but relational ones.

Relational reasons (or reasons of relationality – I'll use these two terms synonymously), as I understand them, fit into a broader conception of *persons as relational,* and of *ethics as involving the character and quality of human relations with one another and with the broader world.* Relational reasons are reasons for action that emerge from our relational connections with and responsibilities to one another. To understand the nature of these reasons, it will be helpful to start by discussing what a relational conception of persons looks like, and to provide some brief background on relational approaches to ethics.

Relational approaches to ethics generally start with a relational rather than an atomistic conception of persons. But what does this mean? Although relational conceptions of personhood are widespread in many ethical traditions worldwide, dominant Enlightenment European approaches to personhood tend to be individualistic. Much of Western ethical theory reflects this focus on autonomous individuals. Social contract theories such as that of Thomas Hobbes (1968; originally published in 1651) presuppose that ethics emerges from rational agreement among independent, self-interested individuals: recognizing that a lawless society leaves everyone badly off, rational individuals enter into agreements with one another to create structures that limit murder, theft, and other harmful or unjust acts. In Hobbes' model, self-interested individuals enter into relation with one another because not doing so leaves them – both individually and collectively – worse off. The system that emerges responds to the question "What kind of system will best advance each person's interests?" and not to "How might we develop a society where people mutually support and relate well to one another?" Social rules exist to help individuals get what they want.

In contrast, a relational conception of persons begins from the assumption that persons are constituted through their relations with one another, and that interdependency is fundamental to the human condition. Confucianism, many sub-Saharan African philosophical traditions, and feminist care ethics all emphasize (in different ways) the idea that our lives and identities are fundamentally shaped by relations with others, so much so that the idea of a fully independent, atomic, autonomous self looks dubious. This doesn't mean that persons are not distinct from one another (it is not as if persons merge into some undifferentiated whole, on these views), but that they are supported, shaped, and become who they are through their relations with others. As Thaddeus Metz and Sarah Clark Miller (2016) put it in their description of the self in relationally oriented ethics:

> [In contrast to] the independent, ideally autonomous, and rational agent who stars in much of modern [European] moral philosophy...[t]he idea of the individual is reconceived as the self-in-relation, a concept that highlights both the fundamentally relational nature of human social ontology, as well as the constitutive importance of relationships for establishing moral agency in the first place.

From this perspective, relationships are central to ethics, and ethics focuses on relating well with others. Put another way, relational ethics emphasize the character and quality of human relations with one another and with the broader world (see Hourdequin 2015, ch. 3; Hourdequin 2021). Acting ethically helps to cultivate good relationships through the development and expression of relational qualities such as respect, care, trust, and reciprocity.

With this background in mind, we can return to climate change. To understand what a relational approach to climate ethics might look like, it is helpful to contrast this approach with one that begins from a more individualist point of view.

4.1. An Individualist Analysis of Climate Change: The Tragedy of the Commons

Climate change is frequently described as a tragedy of the commons (ToC), and despite substantial critique (see, e.g., Gardiner 2011; Patt 2017; Ortmann and Veit 2023), this framing persists, perhaps in large part due to the influential work of Garrett Hardin (1968), who popularized the idea in relation to contemporary environmental problems.

There are many reasons to critique Hardin's analysis, including his application of the analysis to human population and his opposition to basic reproductive rights, particularly for people in poor nations (see Hardin 1968, p. 1246). However, many who reject Hardin's views on population have nevertheless found the ToC framework apt in describing the challenges of climate change. Thus, it is important to consider.

Hardin (1968) illustrates the ToC idea by asking readers to imagine a shared grazing pasture to which each livestock owner can add as many animals as they wish. Hardin argues that in this situation, it is individually rational for each livestock owner to add more and more animals to the shared land, even though the collective effects of such individual choices will result in an overgrazed and unsustainable pasture. In order to avoid tragedy, Hardin argues that some form of regulation is needed, or alternatively, division and privatization of the pasture. (With privatization, which makes each livestock owner responsible for their own section of pasture, the incentive to overgraze diminishes, since the negative effects of overgrazing will cause direct harm the individual owner's livelihood rather than being distributed among users of common grazing land.)

An analogous argument has been made in relation to climate change. In a tragedy of the commons situation where the atmosphere is an open "sink" for greenhouse gas emissions, it is *individually rational* for each person not to limit their emissions, since each person enjoys the direct benefits associated with their own emissions while the costs of such emissions are distributed globally. However, it is *collectively rational* for society to constrain their emissions in order to avoid catastrophic climate change. The trouble is this: even though everyone would be better off if individual emissions were limited and climate catastrophe were averted, it is not advantageous to any *particular* individual to limit their emissions if others are not doing the same.

Baylor Johnson explains further:

> That this is the case is due to the structure of commons problems, which are characterized by three assumptions:

1. The only incentive players have is to maximise [their individual] benefits from use of the commons.
2. The only way players can communicate is by increasing or reducing use of the commons.
3. Use of the commons is shared, [however not all costs and benefits associated with use are shared. Therefore:]
   a. Costs (to the commons) of increased use are shared, but benefits from increased use accrue to the individual ...
   b. Benefits (to the commons) of reduced use are shared, but costs of reduced use are borne by the individual ...
   c. Resources saved by one individual are available for use by any other user.

(Johnson 2003, p. 275)

If these assumptions hold, then individuals in a tragedy of the commons situation will exploit the common resource (in the case of climate change, the shared atmosphere) for their benefit, to the detriment of all.

It is just this kind of description of the climate problem that leads Baylor Johnson – whose argument we considered earlier – to conclude that individuals in a ToC have no obligation to reduce their own emissions, because doing so accomplishes nothing. One person's restraint will be exploited by others: restraint in such circumstances is futile, a fool's errand that only harms oneself.

Johnson does acknowledge that many potential commons tragedies are averted because not all of the assumptions hold. Most people don't care *only* about maximizing benefits for themselves, for example, and people also often communicate with one another through multiple channels, and not only via their increased or decreased use of a common resource. What's more, although Johnson doesn't specifically mention it, more is not always better. The ToC model assumes that each individual has an incentive to consume ever-more of the common resource, without limit. But is it really four times as fun to drive a gas-guzzling SUV 1000 miles on the weekend, as compared to 250? Does flying 100,000 miles a year make someone's life ten times better than if they flew 10,000 miles a year? As the economic principle of diminishing marginal utility suggests, each unit of additional consumption does not have a comparable effect on an individual's wellbeing. In light of this, perhaps the commons might be saved by each person taking what they need (as long as there's enough to go around) rather than taking as much as they possibly can.

A proponent of the ToC model might acknowledge that the assumptions of the model aren't strictly speaking true, but nevertheless insist that they are *true enough* to generate a tragedy. Just look at what has happened over the last century with greenhouse gas emissions! They've gone up, and up, and up. Even among the wealthy, people don't seem to have reached a point of saturation, where their emissions are sufficient to satisfy their desires. Maybe individual restraint in this context *is* a fool's errand. The ToC analysis suggests that a person's well intended emissions-conserving efforts will accomplish nothing.

What's more, perhaps – as Stephen Gardiner argues – the climate situation is even worse than a standard tragedy of the commons. The standard ToC analysis doesn't take account of intergenerational issues, even though climate change involves decades to centuries-long time scales. Because greenhouse gases accumulate in the atmosphere over time and their effects on the climate system are lagged, emissions today may not radically alter climate this year, but instead affect climate years and decades down the road. Similarly, the climatic benefits of emissions cuts are gradual rather than immediate. With this in mind, Gardiner argues that climate change is not a standard tragedy of the commons: instead, it is a complex, intergenerational moral storm, in which each generation may be tempted to postpone action, since the costs of greenhouse gas emissions cuts today will be borne by the present generation while the benefits will accrue primarily in the future. As an individual, one might think, "Why conserve energy? Not only will my emissions cuts be exploited by other people today, any (minuscule) benefits they generate will accrue to future generations rather than to my own!"

In short, according to these analyses, it may look like (1) there is no self-interested reason to conserve (conserving disadvantages oneself), and (2) there is no consequentialist-based reason to conserve (an individual's conservation doesn't actually help climate change), therefore there is no reason to conserve.[3]

But all of this assumes that people are – at base – disconnected and narrowly self-interested, and that people's decisions are driven primarily by the desire to get more for themselves. In a tragedy of the commons framework, those who exercise restraint are viewed as "suckers" whose sacrifices simply leave more for others to exploit. A relational perspective, however, starts in a different place, conceptualizes the problem differently, and arrives at different conclusions.

4.2. A Relational Analysis of Climate Change: Interconnectedness and Relational Reasons

From a relational point of view, the tragedy of the commons analysis of climate change begins in a strange place. Why start by assuming that individuals are disconnected from one another and narrowly self-interested? This is a profoundly *amoral* starting point. It reflects Thomas Hobbes's imagined state of nature, where no rules regulate human behavior, and life is "nasty, brutish, and short" (Hobbes 1968). But why start there? From the perspective of those who view human beings as fundamentally interdependent and as both beginning and continuing their lives in relation with one another, the "state of nature" is not an original condition of humankind, but a degraded one. To land in a true ToC, in this view, means that something must have gone seriously wrong: relations have frayed and broken, trust has been lost, reciprocity has withered, and people have retreated into a narrow and limited sense of themselves.

Although Hobbes seemed to think that "civilization" enables people to progress from a brutish state of nature to a well-governed polity, this teleological view is suspect. The kind of every-person-for-themselves situation Hobbes envisioned seems like an anomalous one, and analogously, true tragedies of the commons may also be historically rare. According to Susan Jane Buck Cox, it is misleading to suggest that unregulated, overgrazed commons are the default, and that commons tragedies abound. From a historical perspective, Cox argues, Hardin's work rests on a misconception of common land in England prior to land division and enclosure in the eighteenth and nineteenth centuries. Before enclosure, English common land was not open, unregulated land, but instead involved specific rights of and limits on use. What's more, contemporary studies of various common resources bear out the idea that communities generally find ways to cooperatively manage share resources and avoid overexploitation, though it may be the case that very large-scale commons – such as the atmosphere, shared by people who are dispersed across the world – are more difficult to manage sustainably due to the scope of the coordination and cooperation needed (Yoder et al. 2022).

Cox suggests that the real tragedy of the commons may be a philosophical one:

> In 1968, Hardin wrote that " 'ruin' is the destruction toward which all men rush, each pursuing his own best interest in a society that believes in the freedom of the commons. Freedom in a common brings ruin to all." But the common is not free and never was free. *Perhaps in the changed perception of the common lies a remedy for ruin.*
>
> (Cox 1985, p. 61, emphasis added)

Although historical commons were rarely free in the sense Hardin describes, the *perception* of commons as a free-for-all can be damaging nevertheless. A shift in perception – from the idea that use of the commons is associated with certain roles and responsibilities, to the idea that the default state of a commons is scramble competition, where each commons user tries to exploit as much of the shared resource as possible for themselves – may be the source of the problem, and the remedy, accordingly, may be a further shift.

Relational approaches, of course, start in a different place, presupposing that we are always already embedded in relationships with others, and that these relationships are sources of moral responsibility. From a relational point of view, people who utilize a shared resource or share common lands, waters, or air, and people who depend on the same ecosystems or the same planet, *are already in interdependent relations with one another, and already embedded in moral relationships with one another.* Ethics – as twentieth-century Japanese philosopher Watsuji Tetsuro put it – is the study of "betweenness," or *aidagara.* Accordingly, the title of Watsuji's book on ethics, *Rinrigaku*, translates as, *"the principles that allow us to live in friendly community."*[4] Watsuji argues that at its core, ethics is about human existence in relation with others. As Robert Carter and Erin McCarthy (2019) explain it, for Watsuji, ethics is about learning "how to navigate these relational waters successfully, appropriately, and with relative ease and assurance," and that the study of ethics is "[t]he study of these relational navigational patterns – between the individual and the family, self and society, as well as one's relationship to the environment."[5]

Relational approaches to ethics therefore "offer a way to consider holistically how to engage well with (human and nonhuman) others, how to interrogate and challenge relations of domination and oppression, and how to repair and reconstruct relations to support individual and collective care and responsibility" (Hourdequin 2021). There are three key ideas important to relational ethics as I am understanding it here:

1. People's lives and identities are developed through their relations with others, and we are embedded in relations of interdependence at multiple scales, from the very local to the global.
2. Living well and developing ethical personhood involves the cultivation of good relations with others; living well does not require achieving autonomy in the sense of independence of others, but rather on a kind of relational autonomy.
3. Responsibilities are relationally dependent (and in some cases, role related).

From a relational perspective, ethics starts with our embeddedness in communities. I might ask myself, to whom am I related, and how? Some of my relations are direct and concrete; others are indirect and obscured by the complex mediation of distance, political boundaries, and international trade, or the way in which greenhouse gases enter the atmosphere and mix together to change temperature, wind, storms, ecologies, and human lives across space and time. Recognizing this embeddedness and locatedness, though – even if some of the relations are like invisible threads – is important. It is from this perspective that one can begin to ask questions about the character and quality of those relations, and to understand how we might enter into better relations with one another, which requires not only attention to dyadic connections between individuals, but deep structural and institutional issues.

With this in mind, what might a relational ethical approach imply with respect to individual action on climate change? What kinds of relational reasons might individuals have for action? Returning to the commons, individuals who share water, land, or atmosphere are not viewed – from a relational perspective – as atomic individuals for whom it is rational to maximize the satisfaction of their individual interests, narrowly construed. Instead, individuals in a commons are connected by their shared dependency and are always already in relation with one another through that dependency.

Reasons for action, on this view, are not grounded solely in a particular individual's desires; they are shaped by everyone's needs and by the overarching goal of reconciling diverse needs in order to develop dynamic harmonies within the community, through which all can be sustained, and through which all can flourish.

4.3. A Broad View of Relational Reasons and its Connection to Climate Change

If ethics is – at least in part – about how to relate well to one another, then we need to consider *all* our relations – relations to those close to us, to those far away, and to both human and nonhuman others. From this point of view, ethics is about is developing, supporting, and sustaining good relations generally, including relations with people one will never meet and with other living beings. Further, relational reasons may involve not only those who exist today but prior and future generations: there can be *intergenerational* relational reasons.

Many ethical traditions emphasize relationality, including feminist, Confucian, and many Indigenous traditions. The phrase "all our relations" translates an idea central to Anishinaabe traditions that emphasizes the importance of the relational connections between people and with all of Creation (McGregor 2009). According to Anishinaabe scholar Deborah McGregor, "environmental justice is about justice for all beings of Creation, not only because threats to their existence threaten ours, but because from an Aboriginal perspective justice among beings of Creation is life affirming" (McGregor 2009, p. 27). This perspective places emphasis on restorative and reparative justice, rather than on retribution (Whiteman 2009). It seeks to avoid negative "relational tipping points" (Whyte 2020) and instead to build trust, responsibility, and accountability.

If ethics fundamentally involves cultivating good relations with human and nonhuman others, and with the land and waters more broadly, then relational reasons are central to ethical life: relations are the threads from which an ethical fabric is woven, and relational reasons are those that support the cultivation of *good* relations.

To make this more concrete, if we think of ourselves as embedded in networks of relations to which ethical responsibilities attach, then *all of our actions matter ethically* insofar as they shape the character and quality of these relations. My profligate burning of fossil fuels can be assessed not solely through a detached calculus of costs and benefits but in terms of the kinds of relations it constitutes or fosters. Perhaps from one perspective I have a *right* to drive a gas-guzzling car for fun every Sunday afternoon if there is no law that prohibits it, but what does this choice say about me as a member of a broader social and ecological community? In the midst of a climate crisis, it is hard to see how this conveys concern for the broader communities of which I am a part, or how my actions help to build the kinds of relations that can enable mutual flourishing. Might I not be able to find a way to spend my Sunday afternoon that would be equally fun or equally meaningful, yet less damaging?

A relational approach encourages us to think about how we are always making and remaking relations with one another and with the broader world through our words and actions. In this sense, this approach is morally demanding and differs from ethical views that focus on individual freedom within minimal ethical guardrails. However, although there are limits to our ability to constitute *all* our relations well within global systems that are structurally problematic, the fundamental approach of relational ethics is likely familiar to many of us in some contexts, at least. In relations with close friends and family, for example, we generally consider how our choices affect these others. Perhaps we have no explicit collective agreement to take turns doing the dishes or sweeping the floor. Nevertheless, helping keep things clean contributes to a dynamic in which people in the household care for one another in ways that support individual and collective flourishing.

Similarly, a relational approach encourages us to think of ourselves as members not only of local communities that include those close to us, but as embedded in broader and larger communities at multiple scales in time and space. As I have already suggested, these communities include not only other human beings, but also diverse plants, animals, and ecosystems. We are also members of communities that are extended in time, including not only present generations, but future generations. How can we orient our lives to live responsibly in relation to the multiple, nested, and overlapping communities of which we are a part? My view is that the answer to this question must consider what each of us *can* do, individually and in cooperation with others, to limit climate change and respond justly to it.

5. Polycentrism, Power, and Possibilities

> To solve climate change in the long run, the day-to-day activities of individuals, families, firms, communities, and governments at multiple levels – particularly those in the more developed world – will need to change substantially. Many of those who need to change, however, have not yet accepted the reality of the threat and their need to act locally in a different manner.
>
> (Ostrom 2009, p. 4)

Climate change is a big problem, and it's easy to feel small in the face of it. It's also easy to think that someone else, somewhere else, should or will solve it. Unfortunately, there's no one person, country, or corporation that can "solve" climate change, and climate change can't be solved in the way a riddle or math problem can be solved. Climate change is – to some degree at least – something we all need to navigate in the decades and centuries to come. We *can* find better ways to respond to climate change, however: we can decarbonize energy systems and reduce emissions to limit warming, and we can find constructive and equitable ways to adapt.

Who is the "we," though? And how do individuals fit in? Each of us is positioned differently in relation to climate change. This means not only that each of us *experiences* climate change differently, but also that we each may have access to different opportunities to effect change. What a person *should* do depends on what they *can* do. For this reason, it can be helpful to consider how actions in different domains can fit together and generate broader synergies. Whether one focuses on reducing one's consumption, developing community adaptation plans, lobbying for political action to accelerate climate mitigation, or promoting sustainable practices in schools and businesses, it can be valuable to recognize that efforts in one domain can support those in another, and that actions at "small scales" – by individuals and families, a small group of colleges and universities, or a particular city or state – can ripple outward, enabling further change. Rather than focusing on a simple dichotomy between individual and collective action, it may be more constructive to think about how diverse actions by a variety of different actors at a range of scales can work together.

This approach to climate action – involving multiple actors and scales – is one advocated by Elinor Ostrom, winner of the Nobel

Prize in Economics for her important research on collective action problems. In a 2009 paper, Ostrom made a case for a polycentric approach to climate change. The term polycentric means multi-centered, and Ostrom uses this term to refer to a multi-centered response to climate change focusing on actions at multiple scales that – ideally – can work synergistically with one another, building momentum for robust responses to mitigate and adapt. Drawing on decades of research on collective action, Ostrom argued that the free rider problems associated with the tragedy of the commons arise at multiple levels. Thus, suggesting that climate change should be addressed by national governments rather than through individual action does not *avoid* free rider problems; instead, it merely shifts them to another level.

Fortunately, collective action problems are solvable and are often solved, in part because the assumptions of rational choice theory that generate the "tragedy" don't apply in many contexts, especially where there are relations of trust and reciprocity (Ostrom 2009a, 2009b). Where people trust each other and consider themselves part of a shared community where there is reciprocity, they don't act only to maximize their own returns, narrowly construed. Instead of acting for their own short-term gain, they tend to act in accordance with social norms that support community wellbeing (see, e.g., Ostrom 2009b).

In addition, relations of trust and reciprocity at local scales can support action on climate change at broader scales. Relations at the micro level, within communities, are critical to macro-level solutions to collective action problems, because laws, regulations, and other government policies function best when there is public support. Imagine if, all of a sudden, everyone stopped believing in the value of traffic lights and simply ignored whether the lights were red, yellow, or green. Even with lots of police attempting to enforce the traffic rules, things would not go well. Put in more general terms, the costs of ensuring compliance are high when people are not aligned with laws, policies, and key social norms. In contrast, when people support systems of coordination and cooperation, relatively little enforcement is needed. In Ostrom's words, "Even governmental policies need to rely to a great extent on willing cooperation by citizens" (Ostrom 2009a, p. 13) and this willing cooperation is developed at multiple scales. In the climate domain, national and local governments can provide incentives for energy efficient appliances or low water landscaping, but trust among neighbors and friends, along with broader commitment to shared goals of resource conservation and climate mitigation, can help these incentives get traction through the choices of individuals and institutions as smaller scales.

There are further reasons to approach climate change at multiple scales. First, international governance processes can be very slow, and policies and institutions at smaller scales may be more nimble, responsive, and innovative. What's more, the actions of entities at small scales can signal changes to come and prompt corporations to take steps to adapt. For example, after California adopted ambitious fuel efficiency standards, the state secured a deal with five automakers to meet those standards, even as the U.S. government under President Donald Trump worked to roll back national fuel efficiency requirements established by the Obama administration. Local and regional scale action can also test out approaches to climate mitigation and adaptation which, if successful, can set precedents for others to emulate. Such initiatives make alternatives to the status quo both thinkable and doable. For years, colleges and universities resisted calls to divest from fossil fuels, citing the risk of lower investment returns. Boards of Trustees often cited these risks as in tension with their fiduciary responsibilities as stewards of institutional resources. But as more and more institutions have begun to divest without negative economic impact, these objections are losing steam. Students, faculty, and staff arguing for divestment in their home institutions can cite data showing that divestment does not carry significant financial downsides (Trinks et al. 2018; Halcoussis and Lowenberg 2019; Plantinga and Scholtens 2021) and point to examples of successful divestment by other institutions. Similar processes can work at the level of individual action: as individuals reduce meat in their diets, for example, they normalize low-meat, vegetarian, and vegan diets. This makes these options more thinkable and potentially more doable for others, as grocery stores advertise meat-free options and restaurants adapt their menus to cater to multiple dietary choices and needs.

5.1. Polycentrism and Individual Action

With the idea of polycentrism on the table, we can consider the insights that polycentric approaches might provide for individual action, and how these approaches connect to the reasons for individual climate action discussed earlier.

Just as relational approaches to climate ethics foreground the embeddedness of individuals in larger networks of social and ecological relations, a polycentric approach encourages us to see ourselves as located in webs of interconnection through which we – as individuals and collectives – have power to effect change. Rather than privileging a single scale or kind of action or creating opposition between individual and collective action, a polycentric approach focuses on how different kinds of action can work together and be mutually supportive. Put another way, *polycentric approaches seek synergies that enable system change.*

In addition, Ostrom's polycentric approach foregrounds the importance of relations that sustain social cooperation. Her work emphasizes the importance of relational qualities such as trust and reciprocity, along with an orientation toward social cooperation, and her polycentric approach suggests that this orientation and the associated relational qualities are built through relationships at multiple scales. Similarly, the relational reasons for action discussed earlier are grounded in the recognition that we are each embedded in communities at multiple scales in time and space, and that paying attention to relational connections at multiple scales can support the goal of developing social institutions, practices, values and norms that enable mutual flourishing.

At a fundamental level, solutions to collective action problems are those that enable mutual flourishing by helping to harmonize diverse interests and needs, and polycentric approaches provide spaces for experimentation and innovation in developing the kinds of relationships and responses that enable mutual flourishing.

Like reasons of relationality, reasons of integrity focus on connections: connections between beliefs and actions, connections among diverse values, and on the harmonious integration of diverse values into one's life. Integrity not only can help to support and sustain individual commitments to action at multiple levels; it can increase individuals' credibility and build trust interpersonally, supporting the kinds of relational connections key to solving collective action problems. Integrity, therefore, is not only an individual virtue that can make one's own life go better; it also enables people to be trustworthy partners in broader efforts to effect social change. What's more, integrity can be cultivated within collectives as well as by individuals: it is not only about what *I* need to do to harmonize various values or align my values with my actions; it is about what *we* can do to find ways to integrate multiple values and develop practices, norms, and institutions that reflect them.

Drawing on Ostrom's polycentric approach to climate change, we can see that reasons of integrity and relationality – although they support climate action – don't support any *particular* set of climate actions on the part of individuals. Individual emissions reductions *can* make a difference, in part in virtue of the fact that we are connected to others and influence others through our own choices, and in part because individual choices add up and can send signals to broader institutions in support of climate-sustainable practices more generally. However, individual action is not only about emissions reductions. There are many opportunities to engage at multiple levels to shift systems toward stronger commitments to climate action with mutual flourishing in mind.

With many possible ways to direct one's individual efforts to respond to climate change, what should each person do? There is no simple formula, but as a starting point, individuals can assess their social location and how that positions them to effect change. For example, students can make an impact by organizing with fellow students to strengthen the climate commitments of educational institutions; engineers can encourage their firms to prioritize sustainable building materials; event organizers can consider how to reduce the impacts of large gatherings through waste reduction and energy conservation; and so on. We don't have to go it alone: we can look around and ask, "What's already happening in my communities? What skills and abilities do I bring to the table? Who can I collaborate with, and how can I use my strengths to contribute constructively to broader social movements to diminish and respond to climate impacts?" We can also find projects with both local and broader benefits. For instance, better systems of public transit can make local transportation more affordable, equitable, and convenient, while also reducing emissions that contribute to climate change globally. Developing water-wise gardens on college campuses in arid zones can add to the aesthetic beauty of the landscape, support pollinators, and reduce pressure on fresh water resources as the climate warms.

By approaching climate change polycentrically and through the lens of relational reasons, we can recognize that each of us has opportunities to cultivate positive relations with others that can support constructive change at multiple scales. Whether through educational institutions, communities of faith, civic organizations, or conversations with friends and family, we all have power to effect change. What's more, those of us who are concerned about climate change can try to integrate our values and actions, and to find ways of living – individually and collectively – that support people's diverse values and needs and sustain flourishing ecological communities.

I'm not saying this is easy. But it's worth a try. Each of us is different: we have different skills, different interests, and different areas in which we are well positioned to make changes. So not everyone needs to do the same thing about climate change. But everyone can do something.

Chapter 2

Opening Statement
Facing Our Limits

Dan C. Shahar

Contents

1. Introduction   47
2. Picking Our Battles   51
3. Dirty Hands   54
4. No Regrets?   60
5. Creating New Norms   64
   5.1. Inefficiency   66
   5.2. Distraction   71
   5.3. Polarization   73
6. Voting Green   76
   6.1. Democratic Duties   76
   6.2. Real Civic Virtue   80
7. What Should Individuals Do about Climate Change?   85
8. Economizing on Altruism   87
9. Conclusion   89

1. Introduction

Before the industrial revolution, most societies saw little change in their living standards from one generation to the next. To be lucky meant avoiding disease, hunger, and violence for long enough to reach adulthood. It meant being healthy and capable enough to labor alongside family and neighbors to secure food, water, shelter, and other necessities. To be especially lucky meant living long enough to meet one's grandchildren. Little could be taken for granted in a world without clean water, electricity, modern medicine, or even a basic understanding of personal hygiene. Life was challenging in the best of circumstances, and it was easily shattered

DOI: 10.4324/9781003146438-3

by famines, epidemics, wars, and natural disasters – of which there were many (Koyama and Rubin 2022).

Even just two centuries ago, roughly 19 of every 20 people lived in what we'd now call "extreme poverty," surviving on no more than the equivalent of $2.70 per day at modern price levels (Roser and Ortiz-Ospina 2019).[1] More than 40% of children died before their fifth birthday (Roser, Ritchie, and Dadonaite 2019), and largely because of this, average life expectancy was just 29 (Roser, Ortiz-Ospina, and Ritchie 2019). Things remained quite bad even in 1950 when more than 70% of people still lived in extreme poverty (Roser and Ortiz-Ospina 2019), more than 20% of children still died before age 5 (Roser, Ritchie, and Dadonaite 2019), and average life expectancy was still under 50 (Roser, Ortiz-Ospina, and Ritchie 2019).

Those of us who live in modern developed societies enjoy a quality of life that would have been unimaginable by virtually any of our ancestors. Even more happily, the proportion of people of whom this is true is growing rapidly. Because of dramatic economic growth over the last few decades, less than 10% of the world population now lives in extreme poverty (Roser and Ortiz-Ospina 2019); 95% of children now live beyond 5 years (Roser, Ritchie, and Dadonaite 2019); and global average life expectancy is over 75 (Roser, Ortiz-Ospina, and Ritchie 2019). There remains a long way to go before the world is rid of poverty, not to mention myriad forms of injustice and oppression. But for the typical person – and, indeed, for the typical member of a disadvantaged group – there's little question the modern era is the best time ever to be alive.

Our prosperity can be traced to numerous developments in technology, commerce, culture, and law (Koyama and Rubin 2022). But some of the most important drivers have revolved around burning fossil fuels for power and transforming nature for agriculture and settlement. These activities have helped lift billions out of poverty. However, they've also had the unintended side effect of increasing the Earth's capacity to absorb and retain the Sun's energy. As a result, our planet is warming at a rate unseen in recorded history. Weather patterns are becoming unstable, oceans are acidifying, and ecosystems are falling into disarray. Scientists refer to these trends collectively as "climate change" (IPCC 2021).

Some commentators have labeled climate change an existential threat to humanity (e.g., United Nations 2018). This is probably an exaggeration. Mainstream assessments project that even without aggressive measures to combat climate change, future people will live better, more prosperous lives on average than people today (e.g., Stern 2006; Tol 2018; for discussion, see Shahar 2019; Heath 2021, ch. 2). However, widespread material prosperity will not prevent climate change from inflicting trillions of dollars of damage on future generations and wreaking havoc on natural ecosystems. Certain coastal and island communities may cease to exist, with their residents scattered to far-flung locations. Numerous plant and animal species will go extinct. In nations with dysfunctional political institutions, people who have benefited little from global development nevertheless will endure climate change's impacts. And all around the world, humans and nonhumans alike will face floods, heat waves, droughts, and many other challenges with greater frequency and severity than before (IPCC 2022a).

The fact we can expect future people to be more prosperous *on average* should not blind us to the serious problems climate change will bring. Our conundrum is that the developmental processes we've historically entrusted with lifting humanity out of poverty are now threatening us, our descendants, and the rest of the natural world.

Fortunately, we're not powerless to prevent these threats from materializing. Nor do we even face a fundamental choice between continued prosperity and environmental integrity. We have the technological know-how to significantly uncouple economic growth from climate impacts and even reverse some past harms. Many of our options for climate stabilization are compatible with preserving the lifestyles industrialization has made possible. We can make considerable progress by developing new power sources, modifying industrial processes, tweaking construction and land-use practices, regrowing forests, adjusting our diets, and pursuing innumerable other measures that – although perhaps inconvenient – would not fundamentally undermine our quality of life (IPCC 2022b).

Yet, on the global stage, it has proven difficult to agree on a path forward. Two of the world's top three greenhouse gas emitters – China and India (Ritchie and Roser 2020)[2] – are also among the countries with the largest populations still living in poverty (Roser and Ortiz-Ospina 2019). Leaders in these nations have been understandably reluctant to adopt measures that might slow the rise of their citizens' living standards or strain their limited governmental capacities. However, without decisive action from them, other nations have been slow to embrace deep cuts. This has been especially true of the United States, where desires for rapid growth are bound up with a longing to remain the world's preeminent superpower. From a technical policy standpoint, geopolitical concerns like these can seem like shortsighted stubbornness. But the fact remains that despite decades of bold rhetoric and posturing, public officials continue to drag their feet on climate change rather than responding with the urgency experts advise (United Nations 2021).

Given the inaction in the policy arena, it's natural to ask what *we* – as individuals – should do. In popular discourse, suggestions abound. Should we restrict our personal "carbon footprints"? Should we donate to environmental nonprofits? Should we become political advocates? In the eyes of many climate activists and environmental ethicists, the answers to questions like these are emphatically, "Yes!" Given the gravity of climate change, these commentators insist people of integrity and principle will respond, even in the absence of coordinated policy solutions (e.g., Jamieson 1992; Hourdequin 2010; Sandler 2010; Raterman 2012; Schwenkenbecher 2014; Hedberg 2018).

Many defenders of a duty to act on climate change qualify this obligation's scope and stringency. For example, Anna Schwenkenbecher grants some individuals may focus on reducing their carbon footprints while others push for collective action (Schwenkenbecher 2014, pp. 182–184). Ty Raterman claims "each individual's moral obligation, roughly, is constantly to strive to do more than she/he does currently and to push her/himself into new, uncomfortable territory, but ... no one is obligated to martyr her/himself for an environmental cause" (Raterman 2012, p. 418). These caveats illuminate that even in the eyes of those who believe individuals must respond unilaterally to climate change, there is room for judgment in determining the specific forms this action might take.

However, in the coming pages, I'll try to persuade you things are even more complicated than this. Against the view each of us is obligated to help tackle climate change, I'll argue that in a world where large-scale public policy coordination is absent, there is no specific duty to take costly unilateral action on this problem. Instead, individuals have broad latitude to choose whether and how to incorporate climate action into their lives alongside other efforts to help make the world a better place. Although action on climate change represents a valid and appealing option for addressing an important global challenge, I'll claim it's possible to live ethically while doing little to combat this problem specifically.[3]

My analysis will have two main punchlines. First, although I deny people are obligated to tackle climate change, I do argue we're obligated to act on social problems in a more general sense. There are many serious issues crying out for our attention, and each of us must decide which of them to pursue and how. The problem with much climate movement rhetoric is that it narrowly interprets a moral duty that's much broader. We all have a duty to help tackle the world's challenges, but climate action is just one of many roles people may take in society's division of moral labor.

The second punchline is that by framing climate change as demanding action from each of us, climate activists worryingly undermine their cause. Given the many challenges we face, it's an empirical as well as a theoretical fact the climate movement must compete with countless other outlets for people's limited attention, energy, and resources. Widespread engagement is not something the climate movement can simply demand; it's an achievement the movement must earn. I'll argue that by taking the need to economize on contributions more seriously, the climate movement can put itself on a better path to success.

My discussion proceeds as follows. Section 2 presents my main argument for the claim individuals have no duty to take costly unilateral action on climate change. As we'll see, this line of argument is vulnerable to a serious objection having to do with the fact we're complicit in helping cause climate change. Section 3 elaborates this complicity objection and shows how it can be repelled. Section 4 explores a different objection having to do with the possibility of "no regrets" climate action. Section 5 considers an alternative line of argument that dodges my initial claims and tries to defend calls for climate action by focusing on social norms rather than independently existing duties. Section 6 discusses whether political action stands out as a uniquely obligatory type of climate action. Section 7 takes stock by returning to our organizing question: What should individuals do about climate change? Section 8 examines my answer's ramifications for the climate movement.

2. Picking Our Battles

Climate change is a critically important problem. Still, it's only one of many problems humanity faces. As I've mentioned already, nearly 1 in 10 humans still live in extreme poverty. Numerous groups face debilitating oppression from their governments, their neighbors, and sometimes even their own families. Many terrible diseases remain without cures, and many people lack access to existing treatments. I could go on, but you get the point. Climate change is a pressing issue, but it's far from the only problem we need to confront.

Just as governments haven't solved the climate crisis, they also haven't solved many other problems. In some cases, it's even governments that are *causing* the problems. And none of this will likely change soon. So, there's pressure on us as individuals to decide what to do. Should we tackle challenges ourselves? Should we donate to promising nonprofits? Should we become political advocates? What combination of responses makes sense in a world with so many unresolved problems?

When we contemplate questions like these, we quickly confront the prospect of tradeoffs.[4] For instance, we have finite cash available to give to charity. We must allocate it recognizing money sent to one organization leaves less to donate to others. Similar things are true about our limited time, energy, attention, and enthusiasm.

We also face tradeoffs between depth and breadth. People who specialize in certain areas typically make disproportionate impacts relative to superficial dabblers. We likely can increase our efficacy by becoming experts on specific causes, developing relevant experience and skills, and cultivating relationships with key players. Unfortunately, this would mean sacrificing our ability to pursue other causes with equal vigor. So, there's a choice to make between doing a little about many problems or doing much more about a few.

Things become even more complicated if we concede we're not prepared to devote ourselves single-mindedly to solving the world's problems. Each of us plausibly has the right to spend time and resources enjoying life with friends and family, pursuing hobbies and interests, and simply taking a break from it all. Yet, carving space for these activities means devoting less of our lives to helping make the world a better place.

Taking these considerations together, it seems reasonable to conclude ethical altruists will not try to tackle every serious problem that comes to their attention. Nor will they always try to act on as many problems as possible. Rather, many will select a small number of issues on which to concentrate, accepting this will mean taking little or no action on many important problems. (For the sake of discussion, I'll set aside whether it's morally *mandatory* to specialize, opting for the weaker claim that this at least seems *permissible*. I find the stronger claim intriguingly plausible, but nothing I say below hangs on it.)

If it's okay for altruists to specialize, that suggests a straightforward argument for the claim we have no specific duty to take costly unilateral action on climate change. The argument goes like this:

1. It is consistent with being an ethical person to concentrate one's altruistic efforts on a small number of problems, taking little or no action on most others.
2. Climate change is just one of many problems we face.
3. Thus, allocating one's efforts to problems other than climate change is consistent with being an ethical person.

The basic idea behind this argument is not that climate change is unimportant or that there's nothing to be said for tackling it. Rather, the point is that important problems are far more numerous than the hours and resources we're obligated to devote to altruism, and we have a right to choose which problems to tackle and how. The diversity of legitimate choices implies there's no *specific* duty to tackle climate change in the way environmentalists commonly suggest. Taking burdensome actions like restricting one's carbon footprint, donating to environmental nonprofits, and engaging in climate advocacy all represent valid options for helping make the world a better place. But they compete with countless alternatives for our limited attention, and individuals can focus elsewhere without doing anything wrong.

This way of thinking about altruism draws an analogy between tackling the world's problems and participating in the social division of labor more generally. Societies need food, medicine, education, and many other things to function effectively. Failure in any of these domains would be catastrophic. Yet, we don't meet our needs by asking each person to do a little farming, a little healing, a little teaching, and so on, in the hope that everyone's small efforts will cumulate in good outcomes. Rather, we recognize the value of specialization for dramatically increasing each person's effectiveness, and we welcome people who are *only* farmers, *only* lawyers, *only* teachers, and so on.

Tackling global challenges differs in important respects from participating in the marketplace. For one thing, we won't always get paid for our efforts – and in fact, it's often the absence of financial rewards that explains why problems are unsolved in the first place.[5] However, the practical wisdom behind specialization and division of labor is just as valid for altruists as it is for profit-seekers. If we embrace this wisdom wholeheartedly, it implies a case for welcoming people who decline to pursue climate action to make time for other issues – again, not because climate change is unimportant, but rather because it's okay for people to focus in diverse ways.

If this argument works, it enables us to see why I said earlier we have no specific duty to take costly unilateral action on climate change. Whereas environmentalists often claim ethical people will feel bound to take such actions, the line of reasoning presented here suggests these calls are more like recommendations to make specific selections from a menu of options. Carbon footprint reductions, donations to environmental organizations, and climate advocacy all are ways we can honor our responsibility to help make the world a better place, but there are other ways to fulfill that duty as well.

It's important to emphasize this argument is not a "get out of jail free" card to justify idleness in the face of climate change. What I've suggested is that because of the limits on our bandwidth for altruism, it's legitimate to focus on a selection of problems while ignoring many others. I've said people who do this have no reason to feel guilty about not tackling climate change despite the fact it's an urgent and serious problem. (By analogy, no one thinks farmers should feel guilty about not practicing medicine despite the fact countless people would die without doctors.) Yet, this line of reasoning won't exonerate people who do little or nothing to advance any important cause. Such people have strong grounds for feeling guilty when someone highlights how little they do about climate change. They also have reason to feel guilty about how little they donate to poverty relief and medical research, how little they volunteer, how little they speak out against injustice, and how little they do to advance an endless list of other causes. I only contend these feelings of guilt don't prove people are required to act on climate change specifically. Climate action is one of many ways to absolve oneself of guilt by helping to make the world a better place.

3. Dirty Hands

Insofar as ethical altruism often will involve specialization, and insofar as tackling climate change is just one way to specialize, the case may seem straightforward for concluding it's okay to decline this specific cause in favor of other options. However, this line of argument is vulnerable to an important objection that targets the second premise I mentioned above, according to which climate change is just one of many problems we face. Although this characterization seems true enough at first glance, one might worry this way of talking conceals an important respect in which climate change is a special problem that demands special responses from us. Specifically, our *involvement* in helping cause climate change may seem to imply we can't discharge our moral duties solely by tackling other problems. To be truly moral, we'd need to act on this specific problem we're complicit in creating.

To see this concern more clearly, consider a stylized scenario. Arielle works on the sixth floor of a large office building. Her office has a large all-in-one printer she must use many times per day. Unfortunately, the printer is prone to malfunction and, on a particularly stressful day, the repeated glitching causes Arielle to snap. She opens the door to a balcony, wheels the printer outside, and tips it over the railing, sending it plummeting to the street below. Fortunately, the printer doesn't hit anyone directly. But as it lands on the sidewalk, it explodes in a cloud of plastic and shattered glass. Some of this debris strikes Brent, who happens to be walking by. Brent is badly hurt and cries out to Arielle for help. Arielle calls back, "I'm sorry, sir! I'd love to help you, but I've decided to focus on other problems instead."

Presumably, we needn't examine Arielle's portfolio of altruistic efforts to know this response won't do. Even if Arielle contributes amply to other worthy causes, this doesn't prove she may ignore any problem she likes. Because she has injured Brent, she must respond to this problem specifically.

Applying similar reasoning to climate change presents a challenge for the previous section's argument. Climate change isn't *merely* one of many problems we face: it's also a problem we're directly implicated in creating. It therefore may seem that just like Arielle can't escape her special duty to address Brent's injuries through altruism in other areas, you and I have a special duty to respond to climate change specifically. Although it *generally* may be true people are entitled to pick and choose which causes to tackle, it's not clear this permission extends to people with dirty hands like us.

Despite its intuitive pull, I believe this objection is mistaken. To begin to see why, let's first ask: does being implicated in creating a harm *always* imply an overriding duty to act on it? It seems possible to come up with counterexamples. Take traffic, for instance.[6] Many people live in cities where traffic is a serious problem: it produces air pollution, stress, accidents, and countless lost hours of productivity and family time. Everyone who commutes in these conditions is complicit in creating the problems. Does this give them a moral duty to take action on traffic?

To sharpen the issue, consider a hypothetical citizen, Chandra, who lives in a traffic-plagued city. Suppose Chandra commutes to and from work every weekday but makes no effort to tackle her community's traffic problems. Instead, let's imagine Chandra is a major donor for her city's women's crisis shelter, an advocate for women and children in city politics, and a volunteer tutor at a struggling school in her neighborhood. Now imagine a local traffic activist approaches Chandra and says, "Don't you realize you're complicit in this city's traffic problem? I mean, I see you're already supporting good causes. But, given how much you use the roads, don't you think you ought to be doing more about traffic?"

I suspect many readers will share my intuition Chandra is not obligated to say yes. Although Chandra's contributions to local traffic may give her some reason to tackle this issue specifically, this seems like the kind of reason that can properly be weighed against competing considerations. For example, could Chandra make a bigger difference by focusing her efforts elsewhere? Is she more passionate about other causes? Do her talents, experiences, and relationships lend themselves better to other forms of action? In our earlier example where Arielle injured Brent, it would have seemed wrong for Arielle to ask questions like these to decide whether to help. But it seems acceptable for Chandra to examine these considerations here and, depending on the answers, potentially decide against tackling traffic.

If you share this intuition, it bears asking: *Why* isn't Chandra required to respond to her city's traffic problems when she's clearly implicated in creating them? What makes this case different from the earlier one where Arielle injured Brent?

Intuitively, it might seem like the answer comes from Chandra's individual insignificance as a source of traffic. Whether or not Chandra drives, her city's traffic problems will be essentially the same. On the other hand, Brent's injuries were due directly to Arielle's actions: if she had left her office printer in its place, the harm would never have happened.

This appeal to causal inefficacy surely is part of the explanation. If every time Chandra got in her car, she caused morally serious harm that wouldn't have occurred otherwise, this would be a powerful reason for her to act. The fact Chandra's driving makes no important causal difference on its own is what allows us even to contemplate the permissibility of focusing her efforts elsewhere.

Yet, this consideration is less conclusive on its own than initially it might appear. To see why, imagine an alternative version of our earlier story where a group of Arielle's colleagues help her destroy the printer in a collective act of vengeance, resulting again in Brent being hurt. In this revised scenario, because enough other people are involved, let's suppose Arielle's individual actions make no difference to Brent's injuries: the same harm would have resulted even if she had stayed inside. Still, the fact Arielle participated in injuring Brent seems to give her a special duty to address this harm.[7] By the same reasoning, emphasizing a single driver's inability to alter the harms traffic causes seems insufficient to validate Chandra's decision to neglect her city's traffic problems.

We can account more successfully for the difference by adding two more considerations. The first is that, unlike endangering people with falling printers, driving is a beneficial activity we don't want to eliminate completely. Although certain people contribute to traffic through actions that are independently objectionable (e.g., driving recklessly, clogging the fast lane, cutting in line to exit the highway), these are the exceptions, not the rule. Halting all driving to prevent congestion wouldn't count as a solution. Part of what we want from a traffic regime is to coordinate travel behavior so people can continue using roadways without collective overuse. By contrast, pushing printers off balconies onto public sidewalks isn't a benign activity we need to keep within safe levels: it's something we want stopped altogether. This makes it sensible to treat Arielle's behavior as a type of *wrongdoing* in a way that's a poor fit for Chandra's driving.

A second consideration is that many of the most promising strategies for alleviating traffic don't revolve around moralized attitudes. For example, it's often possible to reduce congestion by redesigning roadways, changing land use codes, or introducing tolls, carpool lanes, or better transit options. These measures work by making the environment in which people drive more conducive to desirable outcomes. Some do this by altering drivers' incentives to guide them toward different behaviors. For example, introducing a toll on a popular roadway will predictably lead some drivers to decrease use of that route, thereby easing congestion. Other strategies use structural aspects of the driving environment to prevent traffic. For example, replacing stop signs with traffic lights can improve flow without relying on drivers to alter their decision-making habits. For our purposes, what's noteworthy about these strategies is that they function without getting people to view their decisions through a moral lens. This is most obvious in cases involving purely structural solutions that don't rely directly on behavior modifications. But even when incentives are used to alter people's choices, this doesn't necessarily carry the implication that certain responses are morally preferable over others. For instance, even if tolls only alleviate traffic because some people decrease their use of popular routes, people who pay to use toll roads normally aren't regarded as morally worse than people who choose to keep off them. In fact, one of the marks of a successful traffic regime is its ability to accommodate people who place a high value on continuing to use roadways alongside others who can be more easily satisfied with alternative options.

These considerations help us see why it makes sense to regard contributions to traffic as morally different from complicity in causing other kinds of harm. In cases where individuals create problems through inherently undesirable actions, and where moralized attitudes are crucial for solving these problems, it can make sense to demand individualized responses from all involved. But in the case of traffic, pointing to complicity doesn't establish the same type of duty because individual contributions to traffic are benign activities we don't want to eliminate completely and there's no necessary connection between alleviating traffic and turning driving into a moralized issue.

What I want to suggest is that these same points apply to our contributions to climate change. Most individual contributions to climate change are benign activities we don't want to eliminate completely. And there's no necessary connection between alleviating climate change and moralizing individuals' contributions. The same features that explain why it's okay not to become a traffic activist or give up driving in a congested city thus can also explain why our contributions to climate change don't imply a specific duty to act on this problem. Let me expand on these claims in turn.

For the most part, the actions that implicate us in climate change are normal, everyday activities. We contribute to the problem by traveling, working, eating, shopping, and powering our homes, among countless other things. There certainly are some climate-impacting behaviors that are independently objectionable or simply valueless, such that nothing would be lost if people ceased engaging in them altogether. (I'll have more to say about these in the next section.) But more typically, our contributions to climate change come through activities that are beneficial when considered in isolation. It's only because people do these things too frequently or extensively that problems have arisen.

What we want is not to halt these climate-impacting activities completely but rather to find a way to coordinate them to prevent harm. At a collective level, humanity needs to learn how to travel, work, eat, shop, power homes, and facilitate countless other activities in ways that don't add up to a global catastrophe. Yet, when we examine how experts think about achieving this coordination, we see many of the most promising strategies don't revolve around moralizing individual contributions to the problem. By way of illustration, consider this partial list of measures that have been discussed for combatting climate change:

- carbon taxes or cap-and-trade schemes;
- subsidies for lower-impact activities;
- fuel-switching in power plants from coal to natural gas;
- more nuclear and renewable electricity generation;
- expansion of forest cover; and
- capture of carbon dioxide from exhaust streams to store underground.

For our purposes, we needn't focus on the details of each of these approaches. My point in listing them is simply to highlight that none of them revolves around inducing citizens to change their behavior specifically from a sense of moral duty. To think through just one example, take the first item on the list: carbon taxes or cap-and-trade schemes. These are among the most widely discussed measures for tackling climate change, particularly among experts (see, e.g., Nordhaus 2013, ch. 19). Like highway tolls, these policies work by making it costlier to do things that result in greenhouse gas emissions. In the case of carbon taxes, engaging in impactful activities means having to pay a tax; in the case of cap-and-trade schemes, the costs come from needing to purchase credits or forego selling credits one has been allocated. In both cases, the thought is that if impact-intensive activities become costlier, people naturally will do less of them. Crucially, this is not because everyone will think it's wrong to pay the taxes or buy the credits. Like instituting a toll on a highway, these incentive-based policies would imply no moral condemnation of those who chose to bear the added costs rather than changing their behavior to avoid them.

We can say something similar about the other strategies on the list – and many others I didn't mention. Like in the case of traffic, many strategies for tackling climate change operate by manipulating incentives. Taxes and cap-and-trade schemes are examples of this, as are subsidies for low-impact activities. Other strategies are more structural in nature. For example, switching fuel sources in powerplants or incorporating more nuclear or renewables would happen largely at the level of public utilities, not individual consumers. Regrowing forests or capturing greenhouse gases to store underground likewise would affect the atmosphere without requiring morally motivated responses from individuals. In one way or another, each of these measures works to create an environment in which people can avoid a climate catastrophe without moralizing their everyday choices.

The climate problem thus exhibits both features that explained why we should regard contributions to traffic as morally different from other harmful activities. Like with traffic, most people who contribute to climate change are only causing harm because their individually benign actions aren't properly coordinated with others'. Likewise, many of the most promising solutions on the table wouldn't revolve around people changing their behavior from a sense of moral duty. If the analogy to traffic holds, it would imply that although our complicity in creating climate change may provide us with some reason to tackle this problem, this is not the kind of overriding consideration that defeats the case for specialization and division of moral labor I presented in the previous section.

4.   No Regrets?

To this point, I've focused on whether we're obligated to take costly unilateral actions to help fight climate change. Although I grant that we should be ready to accept burdens to help tackle the world's problems, I've also argued we have broad latitude to choose which causes to adopt among our numerous options. However, it may be objected that by focusing so much on the *costs* of climate action, I've exaggerated the extent to which participating in the fight against climate change requires sacrifice (see, e.g., Prinzing 2023). There are many ways to help the planet at little or no personal cost, and some environmentally friendly actions are even beneficial. How should we understand our moral relationship with these "no regrets" responses?

Even if we doubt there's a general duty to adopt costly unilateral action on climate change, it seems plausible we should avoid pointless, wasteful, or self-destructive actions that negatively impact the climate. For instance, it would be hard to defend a person who chose to focus their altruism on problems other than climate change if they decided randomly to set fire to barrels of oil in their backyard. Although basic prudence would also seem to recommend against such senseless behavior, the connection to the climate problem plausibly adds a further moral dimension that makes acts like these *wrong* and not simply *unwise*.

For the sake of discussion, let's grant that in cases where there are *no* drawbacks to actions that serve the fight against climate change, we ought to take those actions. If we accept this, the next question becomes how far the principle extends. It's not very interesting to claim people should avoid senselessly burning oil barrels (for instance) since few people are inclined to do this in the first place. But many environmentalists believe there are wide-ranging opportunities for "no regrets" climate action that involve changing behaviors we engage in regularly. For example, biking more and driving less is both green and healthy, and the same is true of eating more plants and less meat. We can reduce our utility bills as well as our carbon footprints by installing energy-efficient appliances, improving our home insulation, and adjusting our thermostats to avoid overheating and overcooling. Some commentators go even further, arguing a well-lived life typically – and perhaps even necessarily – involves internalizing "environmental virtues" that favor an environmentalist outlook and lifestyle (e.g., Sandler and Cafaro 2005). To cite one example, Joshua Gambrel and Philip Cafaro characterize "material simplicity" as a universal human virtue, claiming most people's unreflecting and profligate consumption is harmful to them even beyond its broader environmental impacts. In their view, people who live happy, virtuous lives will approach purchases more deliberately and less materialistically, resulting in decreased consumption of material goods, greater appreciation of goods that are consumed, and a more focused approach to life in general (Gambrel and Cafaro 2010; see also Andreou 2010).

There's little doubt many of us could improve our lives by adopting some of these recommendations. Many of us already embrace environmentally friendly behaviors in our daily lives and have experienced the benefits ourselves. To the extent there are other such opportunities we have yet to realize, there seems to be little reason to suggest people should feel licensed to deliberately rebuff them.

Even so, there are several reasons to think the scope of "no regrets" climate action is more limited than some would have us believe. One reason is simply that environmentalists routinely ignore or downplay the drawbacks of green options while exaggerating the extent to which they can substitute adequately for more mainstream alternatives. Many consumers have had the experience of an eco-minded friend swearing by a brand of detergent, toothpaste, or other consumer product that turns out to be both expensive and ineffective. Many have listened to dining partners rave about sustainable but inedible food items or held their breath while proud owners of underpowered vehicles attempted harrowing merges into traffic. Of course, green options aren't always disappointing. But dissatisfaction is sufficiently common that many consumers appear to presume that if a product is explicitly marketed as environmentally friendly, its performance likely is worse (see, e.g., Surey et al. 2020).[8]

Just as important are the many intangible costs of altering one's lifestyle, regardless of direction. People rely on habit and experience to navigate their daily lives without having to make endless decisions (Kahneman 2011). Adjusting behavioral patterns can be far from costless, even if the new patterns are no worse in themselves than the old ones. Climate activists often emphasize that once we become accustomed to new ways of living, we'll find them easy to maintain (e.g., Raterman 2012, p. 433), and this undoubtedly is true. But transition costs are real costs, and we can't dismiss them by observing they'll disappear once they've already been paid.

It bears noting that it sometimes can be exhilarating and uplifting to reflect on one's lifestyle and embrace different ways of doing things. We do ourselves a disservice if we let inertia and laziness turn us into knee-jerk reactionaries against trying something new. However, there's no disputing the possibility of erring in the other direction as well. Anyone who has tried adapting to a fundamentally new lifestyle knows it takes lots of experimenting, researching, and figuring stuff out – often more than initially was apparent. Again, these are real costs we cannot simply ignore.

Environmentalists' tendency to downplay these costs also illuminates another limitation on "no regrets" action. When people are passionate about a cause, it's no surprise they display reduced sensitivity to the burdens of promoting it and fixate instead on aspects they find worthwhile. This is a natural response to being immersed in something one finds meaningful and important. Committed environmentalists may not find it especially taxing to inventory household energy use, investigate products' carbon intensity, and undertake many other activities connected to their cause. In fact, they may find such activities positively absorbing. However, what seems easy and beneficial to devoted climate activists may not be equally palatable to people with different tastes and passions. This includes people who focus on different altruistic pursuits and may thus be expected to develop interests linked thematically with their own chosen causes. For example, individuals who focus on disability issues may find themselves more captivated by opportunities to improve accessibility. Individuals who focus on racial inequities may become attuned to improving diversity and inclusion. Once people commit to tackle a particular set of problems, they may find numerous opportunities to act that seem virtually costless. But that doesn't mean these opportunities would be equally costless to people who chose different paths. What constitutes a "no regrets" opportunity often will depend on the specific person considering it.

Reflecting on the many ways people can specialize as altruists supports two main observations. The first is that climate activists do themselves a disservice if they imagine anyone who adopts a lifestyle different from their own must be leading a shallow, selfish, disengaged life filled with empty materialism instead of genuine meaning. Not only is this an intrinsically offensive way to think about people who make different choices, but it also ignores the myriad possibilities for living an excellent life without taking up climate action specifically.[9] When we evaluate green suggestions against the baseline of engaging passionately with other causes, it's easy to see how purportedly costless forms of climate action may be burdensome after all.

However, a second observation is that since many people don't engage adequately with any issues – whether climate-related or otherwise – climate action often will come with "no regrets" simply because the alternative *would* be an existence that is shallow, selfish, disengaged, materialistic, and meaningless. If non-participants tackled other issues in a focused way, the burdensome aspects of climate action would become important as potential reasons not to embrace environmentalists' specific suggestions. But until they begin to do *something*, disengaged individuals will have no good response to climate activists – or, for that matter, any other activists serving a good cause – who propose ways they can begin to help make the world a better place.

5. Creating New Norms

So far, this discussion has focused on *individuals* making choices about whether to participate in the fight against climate change. However, this way of framing the issue runs the risk of underemphasizing how our individual decisions ought to be responsive to the broader social contexts in which we live. In particular, examining personal choices in a vacuum may ignore how social norms can shape and even transform our moral obligations (Bicchieri 2006). Especially in the context of examining social responses to climate change, this is a potentially serious oversight.

To appreciate the transformative power of social norms, let's consider an example from an unrelated domain. In some places, such as Mumbai, India, getting onto a subway can be a physical affair. There often isn't enough room for all who want to ride, and people routinely push to the front of the crowd to ensure they can get onboard. In places like these, people who wish to ride the subway during peak travel hours must be ready to jostle and push like the other passengers. On the other hand, there also are places where very different practices prevail. For instance, in Tokyo, Japan, people in overcrowded subway stations stand in endless lines awaiting their turn to board. Virtually no one jostles or pushes; forcing one's way to the front of the line would leave other riders scandalized.

Now, imagine dropping a person accustomed to Tokyo's practices into a crowded Mumbai subway station and telling them they must board the subway to get home. What should they do? Under these circumstances, it would seem foolish to try to "wait their turn" while the other passengers crowded past. Instead, they should join the fray alongside everyone else. Likewise, a Mumbai resident dropped in a Tokyo subway station would seem well-advised to stand in line.

Given the differences between these practices, it seems clear there's no universal moral truth about how we're obligated to behave when trying to board a subway. What's required of us in a specific subway station depends on the local norms. Even so, these norm-based requirements have genuine moral force. Whereas it's permissible to jostle and push in Mumbai, it's wrong to do so in Tokyo. This illustration helps us appreciate how norms can be *morally transformative*: they have the power to create genuine changes in our moral obligations.

Turning back to the topic of climate change, norms' power to transform the moral landscape creates tantalizing prospects for climate activists (Constantino et al. 2022; Raymond et al. in press). In previous sections, I argued we generally have a right to choose how to allocate our efforts across various causes in our problem-filled world. Yet, even if this is true, it nevertheless may seem possible to shrink our range of discretion by cultivating social norms.

By way of illustration, consider an apparent implication of my earlier arguments: it's permissible to travel by gas-powered car instead of by bicycle, electric vehicle, or public transit (assuming one takes appropriate actions in other domains to help make the world a better place). This conclusion is most intuitive when we consider driving in isolation from the social context in which it occurs. However, imagine if people in our society have generally abandoned gas-powered cars and formed the expectation no one will use them. In this environment, driving a gas-powered car might be regarded as highly transgressive, like forcing ahead of others in a Tokyo subway station. We can imagine similar social developments concerning other impactful behaviors such as travelling by airplane for pleasure, eating meat, and cooling or heating one's house to a completely comfortable temperature. Even if there's no universal moral truth that makes it obligatory to avoid these behaviors,[10] the morally transformative power of social norms could give us genuine obligations to give them up.

To the extent climate activists take themselves to be cultivating new norms, my prior arguments may seem largely beside the point. Perhaps in our current social environment, where strong climate norms are absent, it's hard to establish a specific moral requirement to participate in the fight against climate change. However, it nevertheless could be true that in the type of social environment climate activists are working to create, those moral duties would be very real. Moreover, using moral language to foster new expectations about good and bad behavior can be an important part of establishing new norms. Although such talk of specific moral duties would be technically inaccurate in our current circumstances (because the duties don't yet exist), it could become *self-validating* with sufficient community buy-in. Once enough people embraced new patterns of behavior and expectations, the specific moral duties environmentalists invoke would become real.

As was true in our earlier discussion of "no regrets" action, I don't want to completely reject this line of reasoning. There already are examples of beneficial climate norms making inroads in the population, such as the general expectation to turn off lights and electronics before leaving home. Often, these norms complement the opportunities for "no regrets" action we discussed earlier by targeting wasteful or low-value activities that can be adjusted at little cost. However, the rest of this section will argue climate activists should tread cautiously when treating social norms as a tool to rescue broad individual obligations to act on climate change. Although social norms have the morally transformative power I've described, they also have several drawbacks that undermine their usefulness in the fight against climate change. Specifically, the types of norms most relevant in the climate change context are prone to being inefficient, distracting, and polarizing. When taken together, these drawbacks suggest climate activists will be wise to focus on alternative strategies.[11]

5.1. Inefficiency

Why think it's unwise to tackle climate change by cultivating social norms? Let's start with the claim these norms are prone to inefficiency. This contention has to do with how norms work in practice. In *The Grammar of Society*, Cristina Bicchieri writes, "Norms refer to behavior, to actions over which people have control, and are supported by shared expectations about what should/should not be done in different types of social situations" (Bicchieri 2006, p. 10).[12] Bicchieri's characterization emphasizes that norms work by specifying which actions people are supposed to take or avoid. Norms aren't like personal value systems that guide people's private reflections. They're prescriptions for *behavior* that other people can monitor and enforce.

If we think about how climate activists often think about our duties to participate in fighting climate change, we can easily see this focus on specific favored or discouraged behaviors. As I suggested above, traveling by bicycle, electric vehicle, or public transit can make you a good norm-follower. A person who drives a gas-guzzling pickup truck is doing it wrong. Crucially, we don't need to look inside anybody's mind to know where they stand. If we're wondering whether someone is complying with a norm, we can typically tell by watching.

It's this focus on specific observable behaviors that makes climate norms susceptible to producing needlessly high costs. We can see the problem most clearly by examining an analogous issue. In the domain of environmental policy, "command-and-control" regulations are those that aim to improve environmental outcomes by specifying standards for firms' operations. For example, companies may be made to adopt specific pollution control technologies or limit emissions to specific levels. Well-designed command-and-control regulations can be effective at producing desired environmental outcomes. However, the textbook worry about these regulations is that the standards they set inevitably will fail to be sensitive to the very different circumstances firms face. For any given mandate, it may turn out some firms would face extremely high costs to comply, whereas other firms could achieve even better performance with little difficulty. It's often unrealistic to expect agencies to calibrate their requirements for the details of every individual case (not to mention all the opportunities this can create for corruption and favoritism). Yet, one-size-fits-all policies won't reflect the very different circumstances each firm faces.[13]

The lack of flexibility in command-and-control regulations means it often will be possible to achieve the same environmental outcomes at lower cost by allowing more variation in what each firm does. For example, there could be less of a response from firms facing high compliance costs and more of a response from firms facing low costs. In the end, we could end up with the same environmental outcome with lower burdens overall. (By the same reasoning, we also could choose to impose the same total cost and end up with better environmental outcomes.)

The tendency for command-and-control regulations to produce results inefficiently is why many economists and policy analysts favor incentive-based policies instead (e.g., Mankiw 2021, pp. 193–200). As I noted earlier when discussing carbon taxes and cap-and-trade schemes, incentive-based policies work by making it costlier to engage in impactful activities – but without prohibiting them.[14] This leaves it up to each firm to decide what to do given the specific opportunities and obstacles they face. Polluters finding it difficult to reduce their emissions can either pay a tax or buy permits from other firms. Meanwhile, those who can cut back easily are rewarded for doing so. By making room for a wider range of responses, these arrangements can produce the same environmental outcomes at lower cost (or better environmental outcomes at the same cost).

Returning to the topic at hand, I now want to suggest that if we revise these statements to be about social norms and individual actions, the same basic points will apply. Environmental social norms generate outcomes by specifying standards for individuals' behavior. These standards can't realistically account for the very different circumstances each person faces. Hence, for the same reasons as with command-and-control regulations, these norms will tend to generate unnecessarily high costs relative to the benefits they produce.

We can illustrate this point by returning to the driver of the gas-guzzling pickup truck I described earlier as a norm-violator. To flesh out the example, suppose our hypothetical driver travels 15,000 miles per year in a 2022 Toyota Tacoma. According to the U.S. Department of Energy's Office of Energy Efficiency & Renewable Energy, we can estimate this driver is responsible for 9.9 tons of CO_2 emissions per year (8 tons released directly from the Tacoma's tailpipe and another 1.9 tons released in the production and distribution of its gasoline). Now imagine they'd left the Toyota dealership with a 2022 Prius instead. With its much better fuel economy, driving the same distance in the Prius would have generated just 3.4 tons of CO_2 per year (2.8 tons from its tailpipe and 0.6 tons from producing and distributing its gasoline) (U.S. DOE 2022). So, driving the Tacoma instead of the Prius creates an added 6.5 tons of CO_2 per year.

By treating pickup trucks as transgressive and favoring hybrids instead, climate norms could avert the negative impacts of choices like these. However, because they don't directly account for the differential burdens people face in complying, these gains for the environment can come at very high cost.[15] To get a sense of the magnitude of the implied inefficiency, it's helpful to consider how much people would be expected to pay for the right to emit

CO_2 under a more flexible, incentive-based regime. Policy analysts disagree about what these prices should be, but at the time of this writing, many mainstream commentators cite $50 or $51 per ton as reflecting the "social cost" of CO_2. The $50 number is described by the Environmental Defense Fund as "the most robust and credible figure available," despite EDF's desire to see a higher number reflecting a wider range of climate impacts (Environmental Defense Fund 2021). Meanwhile, $51 per ton was recommended by the Biden Administration's Interagency Working Group on Social Cost of Greenhouse Gases as a tentative estimate for use in U.S. government decision-making (U.S. IWG 2021).

If we accept the higher $51/ton figure at face value, this implies that under a carbon tax regime, our Tacoma driver would need to pay an additional $331.50 per year over what they would have paid with a Prius to maintain their driving patterns. This is not an insignificant figure, and some Tacoma drivers might trade their trucks for more efficient vehicles if they faced this added expense. However, it also seems clear many Tacoma owners gladly would pay $331.50 per year to keep their vehicles instead of switching to more efficient hybrids. Other Tacoma drivers would respond in more moderate ways as well – for example, by keeping their trucks but reducing their driving to avoid some increment of taxation. By comparison to this more flexible, incentive-based outcome, a blanket norm against driving pickup trucks seems strikingly severe.

This impression remains even if we consider much higher price levels for the right to emit CO_2. Some critics argue that if we adopt ambitious goals such as arresting global warming below 1.5°C or reaching net-zero emissions by 2050, a price of $51 per ton is significantly too low (e.g., Stern et al. 2022). However, even if we choose a price far higher than most analysts would consider appropriate, it still looks inefficient to discourage pickup trucks across the board. For example, even at $255/ton (i.e., five times the mainstream figure), our Tacoma driver could preserve their existing behavior by paying $1,657.50 per year. Many pickup truck drivers undoubtedly would balk at paying that much, and these individuals would either switch vehicles or reduce their driving in response. But presumably, there also would be holdouts willing to pay a great deal to protect their vehicular choices. By lumping together all these individuals as transgressors, norms against pickup trucks would deliver environmental results at needlessly high costs.

Although I've focused here on one specific type of impactful behavior, similar issues arise for numerous activities climate norms might target. For example, eating meat, traveling by plane for pleasure, and keeping one's home aggressively climate-controlled are all choices that generate additional impacts over alternative options, and all would become more expensive under an incentive-based regime. However, the marginal impacts of these behaviors are not so great that people universally would prefer to abandon them rather than paying more to keep doing them. By entrenching specific behavioral rules for everyone to follow, a norm-based strategy would produce unnecessary sacrifice on the way to our climate goals.

This problem with climate norms arises because, as we discussed earlier, most of the behaviors that generate climate change are benign when considered on their own. It's because people get real value from these activities that it's worrisome to cultivate blanket norms against them. To achieve our environmental goals efficiently, we need tools that more selectively target activities with low benefit-to-impact ratios. Because norms don't work by considering each person's circumstances, they are prone to being insensitive to situations where their demands are needlessly strict.

Given our discussion of "no regrets" action in the previous section, it's worth pausing to qualify this point. Because some climate-impacting behaviors have very little value to anyone, concerns about treating these activities overly strictly might not arise. For instance, consider the example mentioned earlier of a norm to turn off lights and electronics before leaving the house. Except in unusual cases, people who fail to behave as this norm requires are simply being wasteful. Thus, the argument I've made here provides little reason to object to the norm.[16] Along similar lines, some people may engage in impactful activities only because that's what they perceive as normal and expected, such that shifting social expectations could lead them to change their behavior without much cost (Constantino et al. 2022). In claiming we should be wary of pursuing a norm-based strategy against climate change, I don't mean to imply *every* climate norm is pernicious. My point is that when we look beyond narrow counterexamples to the myriad behaviors that generate climate change, we'll see social norms are prone to being an inefficient tool for producing the outcomes we seek.

5.2. Distraction

Even if social norms aren't the most efficient tool for tackling climate change, this doesn't yield a decisive case against them when better alternatives are hard to come by. As I've said, political leaders have been failing for decades to deliver meaningful action, and desirable incentive-based and structural policies remain confined to expert debates and local experimentation. Against this backdrop, climate norms might appear to serve valuable functions by offering people a way to do *something* about climate change while also fostering public enthusiasm for greater reforms. It is with these thoughts in mind that I offer another reason to be wary of climate norms: in practice, we should expect them to draw attention away from more useful ways individuals, groups, and communities can respond to climate change.

One reason we can expect such a distracting effect is psychological in nature. For some people, the principal attraction of becoming active on climate change is to alleviate feelings of guilt, shame, or distress. For others, the appeal comes from wanting to be seen as virtuous or impressive. These individuals frequently care less about their substantive impacts than about meeting their psychological needs. This in turn may cause them to gravitate toward easy, conspicuous forms of action that soothe their conscience and improve their reputation.

Unfortunately, climate norms threaten to reinforce rather than counteract this tendency. By encouraging people to translate climate concerns into Priuses, solar panels, and reusable bags, climate norms illuminate easy pathways to feeling and looking like someone who's "doing their part." Movement leaders may hope small lifestyle changes will provide stepping-stones to other, more impactful forms of action. But we shouldn't be surprised when many people go no further.

Similar dynamics may arise if people simply have low attention spans for climate action. Given the social pressures that sustain them, climate norms encourage people first and foremost to comply with them. (To recall our earlier example: a person who drives a gas-guzzling pickup truck is a norm violator and can expect to bear the social costs of that violation – regardless of whether they act on climate change in other ways.) For many individuals, we shouldn't be surprised if, by the time they've done all the things climate norms demand, their bandwidth is exhausted for acting on this problem.

Compounding these risks is the fact that opportunities to engage in norm-supported behaviors are tangible, abundant, and easy to comprehend, whereas often it's less straightforward to make progress on climate change in other ways. Which climate-focused non-profits are doing excellent work? Which political candidates stand to drive climate policy debates forward? How can I help my community become better prepared for the future? The climate movement's ultimate success depends on people funneling time, energy, and resources into grappling with such puzzles. However, these are challenging questions without simple, one-size-fits-all answers. By fixing attention on straightforward behaviors one can realistically expect from everyone, climate norms threaten to raise the incentive to take the easy way out.

I've framed these concerns around private individuals' efforts, but the same points also apply at higher levels of activism and politics. Making real progress on climate change often is difficult, requiring reformers to skillfully navigate tradeoffs, uncertainties, and conflicts of interest. By comparison, opportunities abound to engage in norm-supported actions like abandoning gas-guzzling vehicles, installing solar panels, and replacing energy-intensive appliances. If activists and officials can experience the satisfaction and rewards of climate leadership without tackling hard issues, we hardly can be surprised when many do just that (see Brennan 2009).

As an empirical matter, it's difficult to tell how large these distractions have been and are likely to be in the future. On the one hand, we can already see some evidence of "carbon footprint" fetishism and disengagement with other forms of action. Researchers have measured the existence of "conspicuous conservation" whereby attraction to green decisions is driven more by public visibility than substantial environmental impacts (e.g., Griskevicius, Tybur, and Van den Bergh 2010; Sexton and Sexton 2014). Surveys report substantially higher rates of consumer activism on climate change than involvement in political advocacy or volunteerism (e.g., Leiserowitz et al. 2021a, 2021b). Numerous activist organizations and public agencies craft their messaging on climate action in ways that plainly funnel people toward superficial behavioral changes rather than other forms of engagement (often by presenting the former options in compelling detail while gesturing vaguely toward the latter) (e.g., Grantham Institute 2019; UNFCCC 2021;

Earthday.org 2022; Natural Resources Defense Council 2022). And critics of the climate movement have enumerated long lists of high profile but largely symbolic actions against climate change whose costs massively outweigh their atmospheric benefits (e.g., Lomborg 2008).

On the other hand, however, it's hard to know what would be happening if climate activists weren't focusing so much of their energy on cultivating social norms. Would these people who have invested themselves in behavioral changes be devoting their energies toward other forms of action? Would they be doing nothing? The truth is likely somewhere in the middle. However, insofar as we have strong reason to believe moralized behavior changes aren't the most promising way to solve our climate problems – as I argued earlier – we should be wary of activist strategies with such clear potential to distract from what really needs to be done.

5.3. Polarization

I've argued that in addition to being predictably inefficient, efforts to cultivate climate norms risk impeding progress by drawing climate activists away from more promising forms of action. But beyond simply distracting activists, norms also have the potential to create serious obstacles to the climate movement's success by needlessly reinforcing social divisions.

We can see this potential problem by reflecting on the phenomenon of polarization more generally. A robust empirical literature in the social sciences shows most political polarization is not rooted in people's reflective judgments about policy issues: instead, it's fueled by deep divisions in underlying cultural identities (e.g., Haidt 2012; Achen and Bartels 2016). Rarely do citizens enter public deliberations with a truly open mind, examine issues impartially, and gravitate toward candidates and proposals they find meritorious. Instead, they arrive with strong group identities that drive them toward positions that seem congruent with their preexisting self-conception.

These tendencies are evident in climate politics, especially in the United States. As psychologist Dan Kahan writes:

> Positions on climate change have come to signify the kind of person one is. People whose beliefs are at odds with those of the people with whom they share their basic cultural commitments

risk being labeled as weird and obnoxious in the eyes of those on whom they depend for social and financial support.

(Kahan 2012, p. 255; see also
Hart and Nisbet 2012; Kahan 2013)

In some social groups, these pressures push strongly in favor of viewing climate change as a grave problem that demands aggressive action. In others, they push in the opposite direction.

When we consider the behaviors climate norms target, it becomes clear why they have the potential to reinforce cultural divisions. In the preceding discussion, I contrasted norm-following cyclists and Prius owners with deviant pickup truck drivers. Along similar lines, we can point to norm-following vegans and deviant meat-eaters; norm-following hikers and deviant all-terrain vehicle (ATV) riders; and so on. As we run down the list of activities climate norms pick out, we quickly uncover a pattern whereby favored activities are characteristic of one cultural group while those labeled as deviant are characteristic of another. If you're the kind of person who hangs out with cyclists, Prius owners, vegans, and hikers, you'll likely be the kind of person who's already open to climate reform and likewise to the norms we've been discussing. On the other hand, much of the public resistance to aggressive climate reforms comes from the kinds of people who are more at home among pickup truck drivers, meat-eaters, and ATV riders. It would hardly be surprising if members of the latter group regarded climate norms as thinly veiled attacks on their way of life, driving them even further from the environmentalist cause.[17]

It's important to emphasize this tendency to alienate certain cultural groups is not simply an artifact of taking climate change seriously.[18] As I noted earlier, many mainstream strategies for combating climate change would not require people to give up things like driving trucks, eating meat, or riding ATVs. At most, the policies would make these choices somewhat more expensive as people had to bear the full costs of their actions. For those who valued their activities enough to pay the costs, there's no reason even a highly ambitious climate policy regime would have to rule out the lifestyles climate norms would prohibit. Social norms against climate-impacting activities are distinctly liable to polarization because of their inflexible focus on specific favored and disfavored actions.

Stepping back, we now can see why cultivating social norms is a questionable strategy for tackling climate change. These norms offer climate activists a potentially powerful tool because of their ability to change the moral landscape. However, I've argued efforts to cultivate such norms face several important drawbacks. Climate norms are prone to inefficiency, generating high costs to achieve a given reduction in impacts. They are distracting, funneling activists' scarce attention and resources into actions that are unlikely to solve our problems. And they are polarizing, needlessly driving large segments of the population to view the climate movement's efforts as an attack on their way of life.

Together, these concerns help reinforce the position developed in prior sections that we have no duty to take costly unilateral actions against climate change. Independent morality imposes no such duty on us, even though climate change is a serious problem in which each of us is implicated. And although it's sometimes possible to create new duties by cultivating social norms, there are good reasons to question the desirability of employing this strategy here.

The upshot of these arguments is that although each of us has an ethical duty to help tackle the world's problems, morality leaves us significant latitude to choose which problems to tackle and how. We've seen there may be some examples of "no regrets" responses that are appropriate for everyone. But when it comes to more ambitious forms of climate action, I've argued that although these are valid options for fulfilling our duties, it's possible to focus elsewhere without doing anything wrong.

Before moving on, it's worth emphasizing nothing I've said here entails people *shouldn't* change their behavior to participate in the fight against climate change. Green pioneers can serve numerous vital functions in the climate movement, including testing low-impact lifestyles that could eventually become widespread and creating salient precedents for green choices within their community (Constantino et al. 2022). It's true that when behavior modifications become more about making oneself feel or look better than about making an impact (as frequently happens in practice), *that* is grounds for concern. Likewise, climate activists should think twice when their actions generate more costs than benefits, draw attention away from more promising efforts, or inflame social divisions in ways that jeopardize progress. Still, it would go too far to say people *cannot* tackle climate change through lifestyle changes without being thwarted by these obstacles. On the contrary, green behavior modification *can be* a fruitful outlet for altruism. What's unwise is trying to cultivate far-reaching norms that treat these behaviors as obligatory for everyone.

6.   Voting Green

Over the last few sections, I've focused primarily on ways individuals can tackle climate change through costly unilateral behavior changes. Yet, one might object that by focusing on these lifestyle modifications, I've neglected one of the main ways people can advance the climate cause – namely, through political participation. Especially given my emphasis on coordinated policy solutions, it may seem natural to think that even if people aren't required to adopt costly behavior changes in response to the climate problem, we at least can expect everyone to "vote green" (see Maltais 2013). According to this view, a person would fail ethically if they recognized climate change's gravity but didn't translate this concern into action in the voting booth. Even someone who focused on tackling different problems in other areas of their life might seem to have little excuse for omitting this specific action in this specific domain.

This intuition raises two distinct questions for the position I've been defending. First, there is the question of whether individuals who participate in politics are required to do so in a distinctly "green" way. But second, there is a more fundamental question about how we can make sense of a duty to participate in politics at all. Given the arguments I've been making, doesn't voting amount to just one of our many options for helping to make the world a better place?

In this section, I'll address these questions in the reverse order. First, I'll consider whether and how we can understand a duty to participate in politics given the position I've developed so far. Then, I'll examine whether it makes sense to say those who vote are required specifically to "vote green."

6.1.   Democratic Duties

Suppose we grant people have broad latitude to choose how to tackle the world's problems such that taking costly unilateral action against climate change isn't specifically required. On the face of things, political participation would seem to fall into a similar category. Although actions like voting, contacting public officials, and debating others can help advance a variety of important objectives, these are just some of the countless ways we can help make the world a better place. So, it may seem that if we accept my arguments to this point, we should conclude it's also permissible to abstain from politics and focus on other actions instead.

I'll soon explain why I think this view may be onto something. However, I want to begin by considering the opposite stance whereby democratic participation isn't just one of many options for action but rather everyone's moral duty. Many people have the strong intuition that even individuals who engage aggressively with society's problems shouldn't use this as an excuse to neglect politics. When it came time to elect a new president, for example, many would be unwilling to take such people's altruistic efforts as providing justification for avoiding the polls. How can we make sense of this attitude?

The preceding discussion of social norms can help us answer this question. When it comes to democratic behaviors like voting, many of us have strong normative expectations that others will vote, and we believe others expect us to do the same. Moreover, the broad democratic participation linked to these mutual expectations undergirds essential social goods. To the extent our political institutions have legitimacy, this seems linked to their responsiveness to the public's will as it is expressed through democratic processes. And despite their many flaws, there's no doubting democratic civilizations have been among the most successful at fostering peace and prosperity, protecting citizens' rights, and delivering public services without corruption and abuse. (This is not to say democracies always succeed along these dimensions; it's just to say the alternatives typically are even worse.) By sustaining such important values, social norms that urge democratic participation have a powerful claim to our compliance. People who abstain from politics can be characterized reasonably as "free riders" who benefit from others' actions without bearing their fair share of the costs.

We can highlight social norms' centrality to this picture by imagining a society with no established pattern of democratic participation. For instance, picture a country where elections are widely regarded as shams, with little turnout and no legitimacy generated for resulting decisions. If you lived in such a society, it would be implausible to claim you have a duty to show up to vote and even more implausible to call you a "free rider" if you stay home. This is because free riding is possible only amidst a scheme of effective cooperation. (Thinking literally about a "free ride" makes this vivid. Unless there's been sufficient cooperation to produce the train or bus, there's nothing for anyone to "ride" for free.) In democratic societies, it's the actual success of political cooperation – and not any freestanding universal "civic duty" – that puts pressure on each of us to do our part. Without that cooperative success and the mutual expectations at its core, it *would* be legitimate to abstain from politics and focus on other forms of altruism instead.

This helps clarify how we can reconcile a duty to participate in politics with my preceding arguments. In democratic societies, political participation norms produce and sustain essential social goods. By contrast, there's no analogously successful system of norms in the context of climate change. Moreover, it's not clear such norms are what's needed. As we've seen, efforts to cultivate them have significant drawbacks, and there are many other options for tackling climate change. Thus, there is no incompatibility between claiming we have a duty to participate in politics but no analogous duty to engage in climate action.

However, with all of this said, it's also important to acknowledge the existence of serious doubts about whether political participation is morally obligatory. Some authors in the literature even argue it would be better if most people avoided politics altogether (e.g., Brennan 2016; Freiman 2021). The source of these views is the wide gulf between abstract ideals of democracy and the practical realities of politics. Whereas democracy's defenders often wax poetic about convening communities of equals to deliberate over their shared future (e.g., Cohen 1989), real democracy more often manifests as close-minded bickering, empty posturing, and thinly veiled efforts to gain at others' expense. To some extent, these deficiencies can be ascribed to corrupt politicians and special interest groups. However, the broader public also is to blame. Social scientists have shown clearly and repeatedly that most citizens are clannish and ignorant, approaching politics more like sports fandom than like sober deliberation over society's future (e.g., Caplan 2007; Achen and Bartels 2016; Somin 2016).

There is a straightforward explanation for the disappointing quality of most public participation. A typical person's chances of meaningfully affecting political outcomes through voting or civic engagement are vanishingly small. Meanwhile, the costs of becoming a sophisticated, well-informed democratic participant are considerable. As we saw earlier in our discussion of social norms, citizens also face substantial pressures to align with others on whom they depend for social and economic support. So, it's unsurprising people rarely approach politics like dispassionate truth-seekers and take shortcuts instead, embracing oversimplified rhetoric and media narratives that flatter their group identities. As Jason Brennan summarizes:

> If *we*, the electorate, are bad at politics, if we indulge fantasies and delusions, or ignore evidence, then people die. We fight unnecessary wars. We implement bad policies that perpetuate poverty. We overregulate drugs or underregulate carbon pollution. But the problem is that we, the electorate as a whole, don't make choices about whether to be informed or rational about politics. Individuals decide for themselves in light of their individual incentives.
>
> (Brennan 2016, p. 24)

To the extent people participate badly in politics, it's reasonable to ask whether it would be better if they focused instead on other ways of serving their community. Perhaps if these citizens took their civic roles more seriously, it would be desirable to insist they enter the fray. But for citizens whose main contribution to democracy is blind factionalism, the wisdom of pressing a specific duty to participate seems dubious. It might be shrewder to retreat to the more general account of ethical altruism I've described, according to which political engagement is just one of many ways a person can help make the world a better place.

Of course, an alternative reaction to democracy's real-world deficiencies is to seek to improve it, whether by urging better participation from individual citizens or by seeking structural reforms that facilitate better engagement systematically. If successful, such efforts could defuse the concerns I just presented and allow us to embrace norms urging universal participation unreservedly. Yet, when we consider citizens' options for curing democracy's ills, we seem to be pushed even more strongly toward the view I've defended regarding the division of moral labor. Although efforts to fix democracy are laudable, we're not specifically obligated to devote ourselves to repairing our broken politics. It can be entirely legitimate to focus on other causes instead.

In certain respects, these reflections leave us with more questions than answers. To the extent we think mass democracy produces essential social goods despite its deficiencies, this yields a plausible case for embracing participation norms. On the other hand, insofar as we view many citizens' political engagement as (at best) a waste of time or (at worst) a source of social ills, the case looks correspondingly weaker. Thankfully, it's unnecessary for our purposes to settle what position is correct. If there is a duty to participate in politics, I've shown my account can make sense of this by appealing to the transformative power of social norms. On the other hand, if there's no such duty, this leaves political engagement in a similar place as many other forms of altruism: an option for helping make the world a better place that must be pursued carefully to avoid causing more harm than good.

6.2. Real Civic Virtue

Whether or not political participation is morally required, it's appropriate to ask how people should behave if they do participate. Many environmentalists believe the climate crisis implies voters should adopt a principled commitment to support candidates and policies that would tackle the problem. Strikingly, this includes several writers who are skeptical of more expansive individual duties to engage in climate action. For example, Baylor Johnson and Walter Sinnott-Armstrong favor democratic advocacy over other forms of individual engagement because of its power to drive collective action (Johnson 2003; Sinnott-Armstrong 2005). Aaron Maltais adds that "voting green" stands out as specifically obligatory compared to other forms of climate action because its potential for large-scale impact comes at such low cost to individual voters (Maltais 2013).

To the extent we participate in politics, must we do so in a "green" way? In part, the answer would seem to depend on what we mean by this label. Presumably, no one would defend using democratic rights to support candidates and policies intent on ransacking the planet for evil, self-serving ends. So, there are at least some "non-green" political behaviors that seem uncontroversially wrong. On the other hand, there also are interpretations of "green" political behavior that seem clearly optional or even misguided. For instance, it would seem easy to reject the claim we must vote for candidates from a self-styled Green Party regardless of their merits or competitiveness. Likewise, it would seem wrongheaded to insist we always give the climate issue absolute priority over every other consideration.

When we reflect on what it means to be a responsible voter and citizen, it becomes evident there's no simple account of "voting green" we can characterize as universally obligatory. For one thing, in typical elections, citizens must weigh candidates' climate change stances alongside many other factors, including their qualifications and positions on other important issues. Especially in elections where candidates' approaches are not diametrically opposed, it shouldn't surprise us if the person whose climate change platform we favor most isn't the strongest candidate overall.[19]

In addition to accounting for considerations other than climate change, voters also must assess whether a given candidate is likely to deliver concrete improvements rather than simply posturing to gain environmentalist votes (Brennan 2009). Even if we acknowledge the great importance of addressing the climate problem, this hardly implies we must base our voting decisions on hot air and empty promises. And even in cases where we believe a candidate is earnest in their desire to deliver aggressive reforms, it might be wise to temper our enthusiasm if their efforts inevitably would crash amidst opposition and gridlock. When we judge our most preferred outcomes to be politically infeasible, it sometimes makes sense to support less-exciting candidates and policies with a better chance of getting things done.

The details of responsible "green" political engagement become even harder to pin down when we reflect on the complexities of climate policy debates themselves. Even among experts who favor action on climate change, there are important disagreements regarding the best way forward. For example, what role should nuclear power play in our energy future? How should international negotiators handle many countries' inability to credibly monitor, report, and verify reductions in their greenhouse gas emissions? The answers to these and many other technical policy questions are essential for determining which candidates and proposals environmentally minded citizens ought to support.

To illustrate, consider first the question of what role nuclear energy should play in a low-carbon future. Many experts claim nuclear power must undergird any viable strategy to tackle climate change due to its ability to produce vast amounts of low-carbon electricity without the inconstancy of wind or solar generators (e.g., IEA 2019; UNECE 2021). However, this position is controversial among environmentalists because of the potentially catastrophic risks of nuclear plant failures, the insecurity of long-term nuclear waste storage, and the high costs of nuclear construction, among other issues. Those who see risks like these as unacceptable often prefer to focus on reducing greenhouse gas emissions by rapidly expanding electricity generation from renewables like wind and solar (e.g., Lovins and Ramana 2021; Greenpeace 2022). In recent decades, critics of nuclear power have successfully blocked new plant construction throughout the developed world and pushed for the decommissioning of existing plants (e.g., Friends of the Earth 2018; Appunn 2021). Yet, nuclear supporters view this as a misguided gamble on technologies that lack demonstrated success at providing more than a small fraction of society's power (e.g., Shellenberger 2016; Frum 2021).

Each side of this debate characterizes itself as fighting to protect the planet from the other's mistaken and dangerous views. Yet, both sides' assessments cannot be correct. To responsibly exercise "green" citizenship, individuals must determine which view is correct – and what various candidates believe – to avoid throwing their weight behind individuals who would promote damaging agendas once in office.

Consider next the question of how to approach monitoring, reporting, and verification of greenhouse gas emission reductions. This issue rose to prominence in the context of the 2015 Paris Agreement, a global climate treaty famously embraced by the United States under the Barack Obama administration, abandoned by Donald Trump, and then reentered by Joe Biden. Although climate activists have become passionate defenders of this treaty, it was originally viewed as a grave disappointment. This is because, instead of establishing specific duties for tackling climate change, the agreement only requires countries to design their own "ambitious pledges" for action and includes no enforcement mechanism to ensure they follow through. As Bill McKibben wrote at the time:

> In the hot, sodden mess that is our planet as 2015 drags to a close, the pact reached in Paris feels, in a lot of ways, like an ambitious agreement designed for about 1995, when the first conference of parties to the United Nations Framework Convention on Climate Change took place in Berlin.
>
> (McKibben 2015)

The Paris Agreement's largely toothless character was not what was envisioned when the treaty was being negotiated. In the leadup to the United Nations conference that produced it, many hoped the meeting would produce a binding international framework for tackling climate change. However, discussions hit a snag when parties considered how compliance with a treaty would be assessed. Delegates from the United States and other developed nations insisted they could not agree to make aggressive and costly promises regarding their own emissions unless other nations committed to credibly monitor, report, and verify their own impacts – particularly China and India, which together accounted for about a third of global greenhouse gas emissions. Yet, several countries (including China) insisted they lacked the capacity to meet such demands and requested a more lenient arrangement (Clark 2015). In the end, the delegates were able to avert a disastrous collapse of negotiations only by adopting a more modest agreement by which all countries would develop strategies for monitoring, reporting, and verifying their emissions (UNFCCC 2022a) but each country would be given full latitude to determine its own responses to climate change (UNFCCC 2022b). The delegates agreed to return to the table at the end of 2023 with a "global stocktake" of countries' progress, hopefully facilitating more stringent commitments going forward (UNFCCC 2022c).

Did the officials responsible for the Paris Agreement discharge their roles well or poorly? In the eyes of the Obama administration's special envoy, Todd Stern, it was natural to demand robust transparency as a starting point for serious negotiations:

> [W]e have 184 INDCs [Intended Nationally Determined Contributions]; that's great. I mean, it's more than great. It's an absolutely unbelievable start when you think about what anybody could have anticipated. But you need to be able to give everybody in this arena confidence and trust that those INDCs are getting carried out – and "this arena" meaning the countries in the negotiations, but also civil society and observers all over the world – that when the U.S. says it's doing X or China or Brazil or South Africa or Europe are saying they're going to do X, there's a way to track that the X is happening. That's what this is about.
>
> (U.S. Department of State 2015)

Yet, the ineffectual treaty adopted in Paris effectively delayed concerted global action on climate change for nearly a decade – and perhaps more, depending on what happens after the "global stock-take."[20] Here again, it would seem citizens hoping to assess political leaders' "green" credentials must be able to evaluate pivotal decisions like these.

Digging into the details of real-world climate policymaking reveals "green" political participation is not as simple as it might initially appear. Many political actors are eager to represent themselves as friends of the planet, but intelligently distinguishing those whose positions and actions merit support is far from straightforward. Some politicians' environmental rhetoric is little more than lip service aimed at securing votes. Others are systematically ineffective despite making bold, appealing promises. Still others adopt specific policy positions or strategies that are counterproductive or misguided. There even are cases where the officials who are most outspoken about their environmental enthusiasm accomplish less in practice than others who make no such representations. (It's worth recalling that many of the most significant pieces of environmental legislation in United States history – including the National Environmental Policy Act, Clean Air and Water Acts, and Endangered Species Act – were signed into law by the vocally conservative Richard Nixon.)

Does morality require every citizen to navigate these complexities? This would seem quite demanding. Yet, if we don't expect everyone to become climate policy experts, what should we make of the idea of a duty to "vote green"? Should citizens feel required to pledge their allegiance to candidates who engage in empty posturing, naïve overpromising, misguided advocacy, and harmful policy actions, just because those candidates present themselves as "green"? Just as importantly, should citizens feel barred from supporting candidates they consider superior along non-environmental dimensions just because they don't clearly excel along this one?

If "green" political participation involved pressing a magic button that funneled support exclusively to excellent candidates committed to tackling climate change intelligently and efficaciously, the case for insisting upon this would be clear. However, citizens' decisions rarely are as straightforward as this. Few are competent to exercise the level of judgment required to separate the genuine environmental leaders from the dreamers, tail-chasers, and charlatans. And tradeoffs abound between candidates' promises on climate change and their stances on countless other issues.

It would be easy to make too much of these difficulties. Obstacles notwithstanding, it's still reasonable to ask citizens to try their best to account for climate change when exercising their democratic rights. But it seems overly simplistic to suggest honoring this responsibility requires "voting green" in a recognizable and systematic way.

When we connect this conclusion to our prior discussion of a more basic duty to participate in politics, we can see it's difficult to draw a straight line from the climate crisis to a specific individual duty to take political action on this problem. Although democratic governance generates and sustains vital social goods, there are difficult questions to be answered about individuals' duties to participate. And although the climate issue merits consideration by those who exercise their political voice, granting this often will be compatible with not backing the "greenest" candidates and perhaps even supporting individuals who aren't recognizably "green" at all.

7. What Should Individuals Do about Climate Change?

We now can appreciate the difficulty of giving a direct answer to the question of what individuals should do about climate change. Although climate change is a serious problem in which each of us is implicated, it's hard to say what any of us is personally required to do about it. There are many things we *may* do to help tackle the problem, and many of these options are valuable and praiseworthy. A few options stand out as "no regrets" opportunities to help fight climate change at no cost to ourselves, and we presumably should embrace at least these. Yet, because the climate issue competes with countless others for any one person's limited time, energy, and resources, I've argued it's possible to live ethically without taking costly unilateral action in response to this specific problem.

I believe this is the correct analysis of our moral situation as citizens of a warming world. However, I also worry some readers will feel tempted to view this argument as a license for complacency. Some may be drawn to such an interpretation because they desire to escape the burdens of living morally. Others simply will want an excuse to dismiss my arguments out of hand.

But in fact, my position does not license complacency at all. On the contrary, it represents a pointed call to action. Most of us have ample reason to feel guilty and ashamed as we fail to respond to climate change. This is because we don't neglect climate action because we've chosen to focus on other causes; rather, we do it because we're self-centered, lazy, or apathetic. The position I've defended offers no excuse for this kind of behavior.

If my arguments have been correct, each of us faces difficult decisions about how to respond to our problem-filled world. There are far more ways to become engaged than any one person can pursue, especially given the role most of us are willing to grant altruism in our lives. Each of us therefore must choose which issues to tackle and how. Because there are many valid options for making these choices, I've argued there's no specific obligation to select any single cause – even including one as worthy as fighting climate change. But although we have options, we must choose *something*. If we treat the optionality of each alternative as an excuse for choosing *no* alternative – or if we allow ourselves to become paralyzed by the vastness of the possibilities – that is a moral failure.

Many factors merit consideration as we make our decisions. Which problems are most serious and urgent? Where do we stand to make the greatest impact at least cost? What opportunities will go unfulfilled if not by us? Which activities and causes are most likely to hold our interest and enthusiasm? Which forms of action also would serve our personal or career goals?[21] These and many other questions can help determine which options are most promising.

Given climate change's gravity, urgency, and pervasiveness, we shouldn't be surprised if answering these questions leads us to choose this issue as "our cause." Nothing I've said should be interpreted as discouraging such a decision. Saying there are valid alternatives to this path does not imply there's anything wrong with this option. Indeed, the fact you're reading a book on climate ethics gives us at least some evidence this would be a good choice for you.

For those who choose not to take this path, the challenge is to identify worthy alternatives. Here, the danger of becoming paralyzed is very real. Some people will experience paralysis because they let the perfect be the enemy of the good. Before committing to a cause, they'll insist on more research, reflection, and discussion in hopes of finding the ideal way to focus their efforts. In the meantime, they'll let days, months, and even years slip by without doing anything of substance. This is a trap. If acting on climate change – or any other problem – is something we can begin doing *now*, we should consider getting started. We always can decide later to shift to different forms of action. The experiences we gain through early fits and starts only will make us better equipped to make an impact in subsequent roles.

A related trap involves exaggerating the difficulty of taking on new activities and underestimating the ease of adapting to new routines. As I mentioned earlier, it's true that evaluating options, learning new habits, and exercising willpower are costly, and these costs merit consideration. But it's also true that many forms of action are easy, satisfying, and even fun once we overcome our inertia and get started. Although I've largely focused on the need to limit our total burdens as altruists, this should not be interpreted as an endorsement for knee-jerk resistance to new forms of action. We should be open to new options for contributing to valuable causes – whether climate-related or otherwise. We should especially be vigilant for "low-hanging fruit" that allow us to make a positive difference at little or no cost to ourselves.

Together, these considerations help us see why, even in the absence of a specific obligation to tackle climate change, many people nevertheless will sensibly choose this cause. Climate change is a serious problem, and a strong case can be made for discharging one's duty to help make the world a better place by tackling it. Especially in a world where many people don't tackle *any* important problems, climate action offers a salient option for getting started that many people would be wise to adopt.

Still, the fact remains it's possible to legitimately set one's priorities in ways that imply little or no burdensome action on this problem. Our world is full of serious issues calling out for attention. Although climate change is a big, important problem, it's up to each of us to figure out whether to make this "our cause." In this respect, the truest way to begin an answer to what individuals should do about climate change is, "That's up to them."

8. Economizing on Altruism

The possibility of living ethically while declining climate action has important implications for the climate movement. To be effective, climate activists cannot operate from the assumption they have a moral claim to substantial contributions from everyone. Instead, they must accept climate action competes with countless other causes for people's scarce attention. The fact many people don't help tackle climate change is not *only* a result of their ignorance, laziness, and vice: they also have no specific duty to act on this problem. This means climate activists need to think carefully about how they recruit participants as well as how they direct the efforts of those who do participate.

Adopting a "recruitment and retention" mindset would encourage rethinking many climate movement strategies. All too often, climate activists encourage people to adopt behaviors that have little impact in themselves but could produce value if only large numbers of people would buy in, or if only they served as steppingstones for other more impactful actions. In a world where people have no moral duty to participate in climate action *at all*, such strategies present a substantial risk. If people find their experiments fighting climate change ineffectual, they may sensibly choose to redirect their efforts to outlets that promise better "bang for the buck." Such responses would be disastrous for the climate, but they are legitimate and even wise for individuals in a world where opportunities for action dramatically exceed any one person's capacity to act.

For climate organizers, a better approach would focus on offering prospective contributors the most attractive value proposition possible. As I've said, numerous considerations bear on what forms of participation best suit a specific individual. Would-be participants sensibly consider problems' seriousness and urgency, their ability to make significant impacts that would not otherwise have occurred, and the connections between opportunities and their own interests, passions, skills, experiences, and goals, among many other things. Instead of prescribing generic, unenjoyable activities that produce little tangible impact (e.g., reducing energy use, avoiding meat, voting for climate-conscious politicians), climate organizers can focus on creating opportunities for involvement that participants will find stimulating, meaningful, and personally beneficial. For example, it's possible to help fight climate change by planting trees in key urban corridors (Safford et al. 2013), helping neighbors take advantage of public energy efficiency incentives (www.dsireusa.org), and collaborating with local teachers to deliver high-impact educational experiences (U.S. EPA 2022). Unlike many now-popular forms of climate action, activities like these can produce immediate, perceptible impacts that reward those who engage in them, and they have significant inherent appeal that speaks to a variety of interests and aspirations.

Highlighting the diverse ways in which individuals can help fight climate change illuminates a further upshot of the view I've defended, namely: there are many valid ways to participate in the climate movement, and virtually no specific action is required of every participant. As I noted earlier in discussing social norms, many in the climate movement embrace a view according to which responding appropriately to climate change involves adopting specific behaviors and avoiding others. Yet, this mentality is needlessly constricting and likely alienates many people from participating. I've argued there's no reason why someone who, say, drives a pickup truck and loves red meat can't be an earnest contributor to fighting climate change. In fact, the movement's political prospects would improve dramatically if more such people were involved. Yet, climate activists routinely behave as if individuals who don't adopt a large raft of "climate friendly" behaviors are (at best) flawed contributors or (at worst) part of the problem. Embracing and honoring *selective* participation is vital in an arena where individuals can legitimately respond to overbearing demands by taking their efforts elsewhere.

It's worth noting these practical suggestions' value doesn't depend entirely on my arguments' correctness. Even if people had a specific moral obligation to tackle climate change, our empirical reality is one where most people behave as if there is no such duty. It's therefore wise to scrutinize the practical effectiveness of brow-beating people into generic, burdensome, and highly prescriptive forms of action instead of offering a portfolio of enticing options that target a variety of interests, priorities, and identities. The main difference would be that, if there were a duty to act on this problem, climate activists at least would have reason to grumble at having to pander to people's tastes to get them to do what they were obligated to do anyway. If my view is correct, even this grumbling is unwarranted. Individuals have no specific duty to join the fight against climate change. Widespread participation is not something climate activists can simply demand. It's something they need to *achieve* – both in theory and in practice.

9. Conclusion

Climate change is a serious problem that calls for aggressive responses. Unfortunately, such responses have not been forthcoming from the global political arena. Thus, we find ourselves in a predicament where our everyday actions are contributing to a global crisis.

The case for joining the fight against climate change is strong. Yet, our tragic reality is one where we are surrounded by problems that issue powerful calls for action, and we cannot respond effectively to them all. As individuals, we must choose which causes to adopt and which to turn down, recognizing specialization can increase our efficacy at the cost of leaving many problems untouched.

I have argued it's permissible to focus on issues other than climate change and thus to decline costly unilateral responses to this specific problem. By this point, I hope it's clear the emphasis is on *focusing on other issues* rather than *declining responses to this problem*. Because most people take little action on any important problems, the main upshot of my arguments is not that they are excused but rather that they are failing. The key difference between me and my interlocutors is that, in my view, remedying this failure doesn't require tackling climate change specifically. However, it does require tackling *something*.

If I am correct, the climate movement will be wise to pay closer attention to how it recruits and directs people to participate in its cause. Widespread involvement is not something climate organizers have a right to expect for free; it's an achievement that speaks to the movement's success in providing would-be contributors with attractive and meaningful opportunities. Climate organizers also would be wise to acknowledge the validity of many forms of participation. The specific types of responses that have become stereotypical of climate activists *are* stereotypes, and often counterproductive ones at that. Just as there are many legitimate ways to respond to the world's problems, there also are many legitimate ways to participate in the fight against climate change.

If you have read to this point and found my arguments persuasive, then I now invite you to apply them to yourself. Will you help tackle the climate crisis? Will you focus elsewhere instead? The choice is yours. But please don't be just another person who does nothing in a world full of serious problems. Make a decision and follow it through.

Replies

Chapter 3

Reply to Shahar

The Pervasiveness of Climate Change Provides Reasons and Opportunities to Act

Marion Hourdequin

Contents

1. Introduction	93
2. Individuals and Climate Change: Where We Agree	95
3. Individuals and Climate Change: Some Divergences	99
3.1. The Specialization Question	99
3.2. The Tactical Questions	104
4. Conclusion	113

1. Introduction

Does everyone have an obligation to do something about climate change? Climate change is a global and intergenerational challenge. Although sometimes framed as an impending crisis that hasn't quite arrived, climate change is already impacting all parts of the world in diverse ways. For example, in June 2023, New York city was shrouded in smoke from Canadian wildfires, and India was enduring an extreme heat wave. Earlier that year, Australia experienced massive flooding – while drought throughout the Horn of Africa exacerbated food insecurity in Somalia, Ethiopia, and Kenya. Rapid emissions reductions and significant investments in climate adaptation are needed to stem the rising impacts of climate change, and since the problem is distributed – with emissions spread across the globe and adaptation challenges similarly widespread – action is needed on many fronts.

No one person can "solve" climate change, and climate change is better understood as an ongoing challenge to which we must respond than as a puzzle to be neatly solved. Nevertheless, individuals have a role to play in developing constructive responses to

DOI: 10.4324/9781003146438-5

climate change, and many people have the power to make a positive impact. Although individual actions may seem negligible in isolation, they can add up, generating synergistic effects and catalyzing systemic change. In light of these and other considerations, I have argued that those of us who have the capacity to do something about climate change have reasons to act.

In contrast, Dan Shahar suggests that climate action is *optional* for individuals. Although everyone may have obligations to do *something* to make the world a better place, argues Shahar, not everyone has obligations to make the world a better place *through climate activism,* or (especially) through *costly unilateral action.* Climate change is a systemic problem that requires systemic solutions. Some individuals may choose to dedicate their lives to transforming social and economic systems to address climate change, but not *all* of us should feel obligated to do so. Each of us has distinctive interests, skills and talents, and we can be ethically good people if we make positive contributions in some domain, even if it is not the climate domain. Just it is reasonable to leave traffic congestion problems to the traffic engineers and city planners, it is reasonable to leave climate challenges to those with the relevant interest, knowledge, and skills.

Put more briefly, Shahar's central argument – rewritten in my own words – is this:

1. It's (ethically) okay to specialize. Not everyone needs to help solve every problem.
    a. Corollary: it's okay to focus one's efforts on a single problem, or on a small number of problems.
2. Everyone should do *something* to make the world a better place.
3. But climate change is just one of many social and environmental problems we face.
4. Therefore, it's ethically okay to make the world a better place by focusing on problems other than climate change.

Shahar's general argument raises important questions, and in developing his view, he introduces a variety of issues that deserve careful consideration. There are many points on which we agree, and it will be important to highlight those. There are also ways in which our positions diverge, and this chapter describes some of the key differences.

Although it might be simpler if our positions differed more starkly – if Shahar held that no individual has any obligation to do anything about climate change, and if I held that all individuals have significant obligations to act on climate change, regardless of other considerations and contexts – thoughtful disagreements often involve nuance and complexity. What's more, navigating disagreements well requires *attention* to this nuance and complexity. As Shahar points out, there is already plenty of polarization on issues ranging from school curricula to gun laws to the appropriate role for government. Polarization can lead conversations to devolve into shouting matches, or worse. In the United States, we have witnessed this in numerous domains, from Congress to school board meetings – and polarization tends to create acrimony and gridlock rather than understanding and collaboration. In this reply, therefore, I want to illuminate both overlaps and divergences between Shahar's and my views. Our shared hope is that the debate will provoke important conversations and provide resources for others to thoughtfully develop *their* views.

2. Individuals and Climate Change: Where We Agree

So: does *everyone* have an obligation to do *something* about climate change? This question is more complex than it might first appear. To some, the answer might seem obvious. Climate change is a pressing problem to which virtually everyone on the planet contributes, therefore everyone should do something to alleviate it. But with further reflection and some empirical background, we realize that not everyone contributes equally to climate change – in fact, there are massive differences in the average contributions of those who are wealthy as opposed to those who are poor – and the contributions we each make are further mediated by the contexts in which we find ourselves. If I happen to live in a place where my utility company relies primarily on wind and solar power to supply electricity to its customers, my carbon footprint may be much lower than if I live in a place where the electricity supply is powered by coal. People living in these different contexts may use the same amount of electricity, with significantly different impacts. What's more, an individual's ability to act to effect change can vary greatly depending on their skills, abilities, and social position. The mayor of Colorado Springs, where I live, has different capacities to act than I do, based on his role and power. He is much better positioned to catalyze changes in the city's physical infrastructure and public transportation systems than I am. On the other hand, I'm much better positioned to educate college students about climate change and climate justice than he is, since my job centers around undergraduate teaching.

Dan Shahar's position on climate action reflects and builds on these points. Not all of us are well positioned to be climate activists marching in the streets, working on political campaigns, or developing complex policy briefs, and not all of us *want* to do these things. This is reasonable, suggests Shahar. A person who dedicates their life to improving access to health care or fighting for equitable educational opportunities need not feel guilty that they are not also marching in the streets for climate action. They are doing their part to make the world a better place, and that is enough.

I agree with Shahar that not everyone needs to be a climate activist, and certainly not everyone needs to be a full-time climate activist. You can be a good person without being a climate activist or making radical sacrifices to reduce your carbon footprint. I think he's also right that addressing climate change requires structural solutions, and that behavior change can be facilitated by institutions and infrastructure that make it easier and more appealing to "do the right thing." Among those who are able to choose, few people will opt for public transportation if it's more expensive and takes significantly longer than driving an individual car. On the other hand, if public transportation is convenient, affordable, and comfortable, whereas driving is expensive and stressful, and parking is a hassle, more people typically will choose public transit (for discussion of factors that make public transit more attractive, see, e.g., Buehler 2011; Göransson and Andersson 2023). Incentives matter, as Shahar points out, and they can be an important part of an effective climate response.

Incentives, Shahar suggests, are a more effective and efficient tool than social norms that prohibit certain behaviors. For example, shaming people for owning a truck or driving occasionally for fun may backfire by fueling political polarization, and blanket norms can fail to take account of people's individual circumstances. If environmentalism comes with a complex set of lifestyle requirements, it may exclude many. Shahar suggests that people who want trucks should be able to buy trucks and not be shamed for it.

Instead, incentives and costs can be established to encourage energy conservation without imposing blanket restrictions. The incentives-based approach avoids browbeating and holier-than-thou climate activism, Shahar argues, and it allows people greater freedom, while still generating positive change.

I have a somewhat more positive view of social norms and their role in generating constructive change, and although I agree that browbeating is neither desirable as a general approach to generating climate action, certain negative moral emotions – including guilt and shame – may have a role to play. Interestingly, a "name and shame" approach is the explicit strategy associated with the 2015 Paris Agreement, which established the current international framework for emissions reductions. The Paris Agreement requires each participating nation to establish a Nationally Determined Contribution (NDC) – a non-binding climate action plan that includes emissions reduction targets – and these NDCs are reviewed every five years in conjunction with a "Global Stocktake" assessing climate progress. The Paris Agreement seeks to generate progressively more ambitious emissions cuts by requiring countries to share with the global community their climate action plans and progress on these plans, creating accountability through transparency. The idea is that countries that set weak targets or fail to meet more ambitious ones will have to share their progress (or lack thereof) publicly, where they can be praised or denounced by others, reinforcing action and discouraging inaction. It is tricky to assess whether the "name and shame" approach is working in the case of the Paris Agreement, and even if shaming works and is deemed an appropriate strategy in relation to national climate action plans, it may not be the best approach in relation to individual climate action. Thus, Shahar's concerns about a shame-oriented approach are worth taking seriously.

Lastly, and relatedly, Shahar worries that a focus on individual action – especially where it emphasizes lifestyle change – may distract or sap energy from the need for more important systemic change. For example, if I spend lots of time and energy meticulously tracking my carbon footprint and researching which foods and products are climate friendly, this may help to reduce my individual consumption, but do little to effect larger transformation of agricultural production, manufacturing processes, or the energy systems on which we rely. I agree with Shahar that it is unwise to fetishize individual carbon footprints, and that certain changes may not add up to much. We see this not only at the individual level, but also with corporations, nations, and other large-scale actors. Modest changes are sometimes announced and celebrated with great fanfare even when they fall tremendously short of what is needed to reach the targets required to keep warming under 2°C. The efforts and actions we collectively take to address climate change should be proportionate to the scale of the challenge; taking modest or misguided steps that amount to mere window dressing won't do.

Shahar's core argument focuses on the specialization question: as long as we each do something to make the world a better place, is it okay for some of us to ignore climate change when it comes to individual action? I'll address this question in further depth below, because our different answers to it reflect somewhat different conceptualizations of climate change and divergences in our conceptions of ethics. There are some fundamental theoretical differences that will emerge through this discussion.

Many of Shahar's further points explore tactical questions about how climate action is best approached, and what the answers imply for individuals considering what they ought to do. For example, to what extent are different forms of individual action actually effective in moving the needle on climate change? Do individual emissions reductions distract from or add momentum to larger efforts directed toward systemic change? Does shaming work as a form of moral motivation, or is it more likely to backfire, leading people to turn *against* climate action? Does everyone have an obligation to at least "vote green" (Maltais 2013)?

I agree with Shahar – up to a point – on the specialization question, but because I see the challenge of climate change as intertwined with virtually all aspects of everyday life for future and present generations of humans, and with ecological systems across the planet, I believe that we simply can't avoid being involved as individuals – in one way or another – in shaping climate futures, and that this entanglement carries with it ethical obligations (or as I prefer to put it, certain ethical reasons for action).

The tactical questions are more multifarious and in many cases hinge on guesses about how people and institutions will respond in various contexts. Answers to tactical questions can be tricky to decisively determine. We can draw on empirical research for certain insights, then try to evaluate how specific approaches work in specific contexts (e.g. Is naming and shaming working to increase individual countries' climate pledges and emissions reductions efforts?

Does shaming prompt individual climate action?). Nevertheless, it is unlikely that these tactical questions will be settled once and for all, since the tactics that will be most effective in a given domain often depend on the particularities of the circumstances.

In this section, I have highlighted some of the important insights that emerge from Shahar's arguments, and some of the places where we agree. In the next section, I explore in more detail the ways in which our positions diverge.

3. Individuals and Climate Change: Some Divergences

3.1. The Specialization Question

Shahar argues that climate change is just one of many social and environmental problems we face, and it's not reasonable to expect everyone to do something about everything. Thus, although everyone should do something about *something*, not everyone needs to do something about *climate change*.

As I have said, I agree with Shahar up to a point. Not everyone needs to be equally engaged in addressing climate change, and not everyone needs to do the same thing about climate change. Put in terms of reasons, not everyone has reasons to respond in the same way or to be equally involved in addressing climate change. On the other hand, I do think that almost everyone has reasons to do something about climate change, and those who are particularly concerned about climate change and who place value on ameliorating climate impacts have additional reasons to do something, based on these values and concerns.

Shahar might concede that those who are actively concerned about climate change have reason to do something about it, and yet insist that not everyone needs to be concerned about climate change. The specialization argument suggests that it's okay for people to focus their attention, action – and presumably, concern – on certain issues, to the exclusion of others. As Shahar reminds us, climate change is just one of many problems we face.

Here is a place where we differ. In my view, climate change is not just one of the many problems we face. Because climate change is so wide ranging and significant in its effects, it is intertwined with and in many ways a fundamental component of all – or almost all – other problems we face. Take the problem of global hunger. Although there is theoretically ample food to support the full human population globally, there are distributional challenges, and these distributional challenges are significantly exacerbated by climate change. Climate change poses a significant threat to food security worldwide (Dasgupta and Robinson 2022). Or consider the dependence of global commerce on complex computer systems and "cloud" storage: this, too, is entangled with climate change, because the servers that support companies and institutions worldwide consume huge quantities of energy, much of which is produced through the burning of fossil fuels (Singh et al. 2021; Monserrate 2022). Here is a third example: in 2022 and 2023, two major U.S. insurance companies – Allstate and State Farm – announced that they will no longer offer homeowners insurance in the state of California due to increasing climate-related risks (Mac 2023; State Farm 2023). Major insurance companies have also withdrawn from Florida and Louisiana (Flavelle 2023), and insurers are pulling back from numerous areas across the country deemed at high risk due to flooding, increasingly severe storms, or wildfires. Both directly and indirectly, climate change is having significant impacts on where people are – or are not – able to live. In virtually every sector and institution – including state and local governments, the news media, school systems, banks, corporations large and small, farms and ranches, religious communities, and beyond – people make decisions that can ameliorate or exacerbate climate change and its impacts. Schools can choose to incorporate or exclude climate change from the curriculum. Banks can support or divest from fossil fuels. State and local governments can develop climate adaptation plans and climate-smart infrastructure. Farms can adapt irrigation practices to respond to warming climates. Corporations can shift toward renewable energy sources and take numerous other steps to adopt more sustainable practices.

Not everyone is a corporate CEO, a college president, or a member of congress, and many of us lack the power associated with positions like these. But most of us do have *some* power to effect change in certain domains, and because climate change is such a critical issue, we have reason to take it into account and to identify ways in which we can contribute – whether in ways small or large – to ameliorating climate change and its impacts. For those who consume resources profligately, this might involve lifestyle changes to reduce air travel, keep clothes for longer or curtail meat consumption, but lifestyle changes are far from the only way to contribute. College students have had significant impacts by lobbying for campus divestment from fossil fuels (for a review of fossil fuel divestment in U.S. higher education, see Barron et al. 2023). Many lawyers do *pro bono* work, some of which could focus on climate-related cases. Engineers can consider energy-efficiency in their design practices, and develop technologies that support the transition away from fossil fuels. In my own field, a small group of concerned philosophers founded a new group, Philosophers for Sustainability; developed a sustainability best practices guide; and encouraged the American Philosophical Association to adopt these practices, which it did. Efforts in each domain may be modest, but they can add up.

Specialization operates here *within* the realm of climate action. Many people have opportunities – within the context of their everyday lives – to contribute positively to constructive climate responses, even without becoming single-mindedly focused on climate change. Because climate change is such a serious and all-encompassing challenge, it can't be left to specialists in a single domain. Climate modelers can help generate forecasts of climate impacts, but they aren't the ones who are well positioned to transform energy systems or develop adaptation plans. Engineers can design increasingly efficient photovoltaic panels, but others are needed to manufacture and install them. Activists can march for more ambitious emissions reductions targets, but policymakers need to figure out how to implement these targets.

Given the wide range of responses required to adequately address climate change, we need an all-hands-on-deck approach. Luckily, this doesn't mean that most people need to become climate activists; many of us can become climate *advocates* – to some extent – by simply integrating concern for climate change into what we already do.

Shahar's view supports specialization at a different level, where only some people need to work to address climate change, while others focus their efforts elsewhere. Our views converge to a point, because I agree that not everyone needs to give climate change the same level of attention or respond similarly to it. However, in my view most people have reasons to consider climate change as they make choices as individuals and as members of various institutions and communities.

Both views are susceptible to the objection that our positions lead to slippery slopes. Shahar insists that not everyone has an obligation to do something about climate change; however, everyone has an obligation to address *some* social or environmental issue. But why? If specialization is okay in relation to climate change, why not in relation to making the world a better place more broadly? Extending Shahar's logic, isn't okay for some people to be social changemakers, and for others of us just to go to work, feed our kids, and have some fun in our free time? Shahar could, perhaps, reply that just by doing these things – being responsible employees or taking care of those close to us, for example – we are making the world a better place. However, that is not how I interpret his argument. Just humming along with daily life without a thought to the significant social and environmental issues we face is, according to Shahar, a problem. So the questions that arise are these: Why do we *all* have to do something to make the world a better place? Can't we outsource this responsibility to those who are good at it, leaving the rest of us free simply to live our lives as we see fit without worrying about "saving the world"? Maybe just a few superheroes can do that work on our behalf (that's how it goes in the movies, anyway!). If Shahar's specialization argument implies that people lack individual responsibilities to do something about climate change, then perhaps his argument can be extended to imply that we don't have responsibilities to do something about *any* social or environmental issue. There may be a slippery slope from specialization regarding *which issue to address* to specialization regarding *who addresses social and environmental issues* in the first place.

Turning to my view, there are distinct, but analogous slippery slope worries. If climate-related responsibilities – or at least climate-related reasons for action – are more expansive than Shahar believes, and if most people have reason to do something about climate change, then why don't most people have reason to do something about *all* social and environmental issues? If we're on the hook for climate change, what about global poverty? Or violence and war? Or inequitable housing practices? What, if anything, makes climate change "special"?

I have already offered at least a partial answer to this question. Climate change is distinctive because it is pervasive and shapes life prospects for virtually all inhabitants of the planet (human and otherwise) for decades and centuries to come. It is a fundamental challenge to the flourishing of humans, nonhuman animals, plants and ecological systems on Earth. What's more, most people not only contribute to climate change but are capable of doing something constructive in response. Although it is true that we need systemic social, economic, technological and political change in order to address climate change, systems don't just change themselves, so as participants in a variety of institutions and systems, we can ask ourselves where we have access to levers for change and pull on those levers.

Climate change is a distinctive challenge in many ways, but the arguments I have offered may apply to other social and environmental problems as well. Ethical specialization, in my view, only goes so far. It might be the case that I don't have reasons to do anything to address a minor chemical spill in a part of the world I have never visited and to which my own actions have little connection, but that doesn't mean I have ethical reasons for action only in relation to issues that I *choose*. Just as I have basic obligations to respect others that can't be outsourced to ethical saints, I have reasons to help ameliorate certain pervasive social problems in which I am implicated to a greater or lesser degree. For example, I believe it is the case that everyone who is capable has reasons to act to diminish racism and other forms of oppression, even if they were born into social worlds in which racist institutions already existed. These reasons for action may take different forms depending on one's social positionality, and they don't imply that every person needs to be a full-time civil rights activist. As is the case with climate change, we can find opportunities to act in the communities and institutions in which we find ourselves.

Shahar's view on specialization emerges in part from his conception of what is required to be an ethical person, and the associated, implied concern that the requirements not be overly demanding. Shahar suggests that doing something to make the world better is required to be an ethical person, although doing something about climate change specifically is not. I'm disinclined to put the issue in these terms. I agree with Shahar that it's not especially helpful to label people who aren't doing anything about climate change as unethical, but Shahar's framing has the potential to make ethics sound very binary, like there's a sharp threshold or set of necessary conditions for being an ethical person. In my view, ethics is about relating well to others and the broader world, and in many contexts, there aren't clear boundaries or check boxes for this: our responsibilities are often open-ended. It's not as if I can reasonably say, "I was nice to three people this morning, so it's fine if I'm rude this afternoon, since I've done my part." Relationally oriented ethical approaches suggest that relating well to others and to the broader world is something that is woven into the fabric of our lives and interactions. In this sense, we're always "on duty" from an ethical perspective – not in a way that demands constant self-sacrifice, but in a way that suggests we be generally thoughtful in how we live our lives and relate to others. Thus, although I don't deny that it may make sense to specify ethical minima in certain domains (honoring basic human rights protections comes to mind), I'm disinclined to think about individual obligations in relation to social and environmental issues as involving certain minimum requirements to qualify as an ethical person. Although the absence of clear minimum requirements to be an ethical person may make it harder to know when one has done enough, it may also moderate worries about over-demandingness. Although almost everyone has reasons to do something about climate change, there's no specific amount of action required to be an ethical person, on my view: each of us should do what we reasonably can, and precisely what that looks like may vary quite a bit from person to person.

3.2. The Tactical Questions

When considering what individuals should do to address climate change, a number of questions arise that are – at least in part – tactical. They focus on the extent to which individual actions contribute – or fail to contribute – to constructive climate action, to what extent moralizing individual action is effective and desirable, and what some of the potential side effects of a focus on individual action might be. Put another way, what are the consequences, whether good or bad, of focusing on individual obligations in relation to climate change? More bluntly: Can individual obligations save us from climate change? Can they get the job done?

After offering his core argument for ethical specialization, Shahar further defends the view by considering and replying to some potential objections, and by critiquing approaches that emphasize individual obligations. These objections and critiques focus primarily on the potential *consequences* of approaching climate change through an emphasis on individual obligations. If individual climate action has few significant positive impacts, then this may support the specialization argument by suggesting that individuals focus their efforts elsewhere. Shahar questions the efficacy of individual action and efforts to promote such action through the promulgation of norms, and he questions the role of individual action in systemic change. But first, he considers a more fundamental objection to the specialization view: the idea that individuals ought to do something about climate change because they have helped to cause it.

The Dirty Hands Objection

The first objection to Shahar's specialization view focuses on "dirty hands." According to this objection, climate change can't be considered just one of a range of potential issues that individuals might choose to address because individuals actively *contribute* to climate change. The underlying intuition is that we should take responsibility for and clean up our own messes, or make amends for harms that we cause. Imagine that I accidentally throw a baseball through your window and decline to pay for it, saying that your broken window is just one of the world's many problems, and I'm focusing instead on plastic pollution, thereby doing my part. This rationale surely won't fly. I should apologize and pay to fix the broken window.

Shahar acknowledges the pull of the "clean up your own mess" intuition but suggests that its applicability depends on the context. Two examples illustrate these contextual considerations. Shahar's first example involves a frustrated employee dealing with a malfunctioning printer. If, in a fit of rage, this employee throws a giant printer out the window of their office building, harming pedestrians below, then the printer-tosser is morally responsible for the harm they cause, which is a direct result of their action, and they should make amends. On the other hand, if someone drives to work and in doing so contributes to traffic congestion, Shahar thinks that they don't thereby have an obligation to help resolve the city's traffic problems. That can be left to others.

Shahar's reasoning about the printer case focuses primarily on the intuition that drives the dirty-hands objection: those who contribute to a problem should be the ones to fix it. The printer-pusher directly and negligently causes harm to people below, and this person can't simply beg off, saying that others (specialists in remediating the harms caused by angry employees pushing printers out of office windows, perhaps) can address the damage.

Shahar views the traffic case differently, however, and his response to this case mixes ethical and tactical considerations. He argues, first, that a single individual's contributions to a city's traffic problems are negligible, so causal responsibility looks different here than in the printer case. More importantly, Shahar suggests, driving to work is not in itself a bad thing (whereas pushing printers out of the window of an office building presumably is). People need to get places to do their jobs, and driving is a way to get there. In addition, Shahar argues, making people feel guilty for driving to work won't accomplish much. Traffic problems need structural solutions such as incentives not to drive at peak times or synchronization of traffic lights to improve the flow and prevent backups.

What Shahar suggests, then, is that an individual's relation to climate change is more like the driver's relationship to traffic problems than the printer-pusher's relationship to the injured pedestrians. The printer-pusher has both a moral obligation to help the people they injured and the ability to help by calling 911, offering first aid, paying the injured people's medical bills, or the like. In contrast, the driver's moral obligation is attenuated by their relatively minor part in the problem, the positive value of driving to work, and the need for structural rather than individual changes to address traffic issues.

I agree with Shahar that the second case provides a better analogy to individuals' relationships to climate change, but for two reasons, I'm not convinced that the considerations on offer imply that individuals lack ethical reasons to act. For one, we *do* contribute to climate change, as Shahar acknowledges, and even if the contributions of a single individual are modest in relation to the scale of the problem, these causal contributions entangle us with the issue. Climate change is not something we can claim to have no part in or to be utterly disconnected from. On the other hand, it is true that each individual – even those who are profligate emitters – contributes just a small slice of total emissions, so no one individual should have to bear the full moral weight of responsibility for climate change, but rather a small part of it.

A second reason why I'm not fully convinced that the traffic example is sufficient to show that individuals lack strong reasons to act in relation to climate change is because climate change requires a very multi-faceted, multi-level, multi-sector response. Even if traffic problems can be solved by traffic engineers, climate challenges can't be solved by a single set of specialists. It takes a village. We're all implicated in climate change through our contributions, but it's not only our dirty hands that give us reason to help: insofar as we have the capacity to do something about a challenge that poses a major threat to the flourishing of human and nonhuman life now and for centuries to come, we have reasons to act. This of course raises the question of *how much each of us is required to do* and to what extent we should make sacrifices to address climate change. And that brings us to the issue of low hanging fruit.

The Low-Hanging Fruit (or "No Regrets") Objection

After considering the dirty hands objection, Shahar turns to the question of low-hanging fruit. Can individuals engage in climate action at little to no cost to themselves (and perhaps even with some personal benefit)? If so, this weakens the specialization argument, since individual climate action can easily be woven into people's everyday lives and choices. But Shahar doubts that there's much low-hanging fruit in the climate arena. He describes efficient but dangerously underpowered cars and ineffective "green" laundry detergents as examples of the false promises of so-called low-hanging fruit. I don't think the situation is nearly as bleak as Shahar suggests in relation to individual lifestyle choices; however, the accessibility of climate-friendly options in many cases depends on one's context and resources. For example, electric cars are both efficient and powerful, but their up-front costs are significantly higher than their gasoline-powered counterparts. Although the *lifetime* cost of owning an electric vehicle (EV) may be on par with vehicles that rely on gasoline (Weldon, Morrissey, and O'Mahony 2018; Liu et al. 2021; for broader discussion of the economics of electric vehicles, see Rapson and Muehlegger 2023), this may vary based on contextual factors such as initial cost, miles driven, and tax incentives or disincentives (see, e.g., Malima and Moyo 2023; Kumar, Kalghatgi, and Agarwal 2023). What's more, even in wealthy countries, many people lack the financial resources to invest in an electric car due to the significant initial cost.

Although I do think Shahar is unduly pessimistic about low-hanging fruit in relation to personal emissions reductions, his focus on carbon footprints may be the main source of divergence in our views of low-hanging fruit. For reasons described in Chapter 1, I believe people have reasons to reduce their individual emissions, but this is far from the only way to contribute to climate action.

If one takes the approach I have suggested – taking advantage of one's knowledge, skills, and abilities to effect change in the institutions and communities in which one finds oneself – it is possible to find low-hanging fruit in these contexts, not only in the realm of personal emissions. For example, as a college professor, with relatively modest effort, I can support students advocating for divestment of the college's endowment from fossil fuels; I can push for more sustainable practices in my department; and I can develop courses that focus on climate ethics and climate justice. Students can join a campus environmental club, engage in community work focused on climate resilient communities, intern for the office of sustainability, take classes to learn more about climate change, or simply support waste reduction and energy conservation in the dorms. People who attend religious services regularly can engage in conversations with their faith communities about what they can do to address climate change, and workers in various contexts can suggest minor changes that reduce waste and conserve energy – saving money as well as reducing climate impacts. There is no one-size-fits-all recipe for individual climate action, and it is important to recognize this. Thus, I agree fully with Shahar's point that committed climate activists "do themselves a disservice if they imagine anyone who adopts a lifestyle different from their own ...[leads] a shallow, selfish, disengaged life filled with empty materialism" (p. 63, this volume) – but the view that most individuals have reason to do something about climate change needn't lead down this problematic path.

It is true that climate change requires significant institutional and systemic changes, and not all fruit is low hanging – but *some* is, and there are plenty of opportunities to harvest that fruit. Many forms of individual climate action are not onerous, and it is possible to act on climate change without becoming a full-time climate activist. Thus, individual climate action is compatible with specialization in many other domains.

Norms versus Incentives

Both the dirty hands and low-hanging fruit objections involve questions about what's needed, what works, and what's feasible when it comes to climate action. Shahar's concerns about the role of norms in responding to climate change raise related tactical questions:

- Can changing social norms catalyze the social and institutional transformations needed to address climate change, or are incentives a better approach?
- Can norms unite people in common cause, or are they more likely to generate polarization?
- Do norms that guide individual behavior distract from effective climate action, or can they help to catalyze it?

On the first question, Shahar suggests that incentives are much more efficient than norms, because incentives allow people (and institutions) greater flexibility in their choices. Do you love driving a truck? That's not the lowest impact choice, but there's no need for moral judgment: incentives can be established so that pickup truck drivers pay the social cost of the extra carbon they emit. Love to travel? That's fine, you just need to pay a bit more to account for the carbon-intensive costs of flying. Shahar's point is that incentives (and disincentives) give people the agency to decide what matters to them, and to act accordingly. Yes, they need to pay more for carbon-intensive activities, but that's up to them. If they realize that they don't value driving a Toyota Tacoma that much, maybe they'll choose a more efficient car and save the carbon tax.

That all sounds reasonable, but in certain cases incentives can backfire, undermining existing moral motivations. For example, economist Samuel Bowles offers an example from a daycare center in Haifa where instituting financial penalties for picking up children late actually led to parents' *increased* tardiness in collecting children at the end of the day. Bowles suggests that "[t]he fine seems to have undermined the parents' sense of ethical obligation to avoid inconveniencing the teachers and led them to think of lateness as just another commodity they could purchase" (Bowles 2008, p. 1605). Drawing on this and other empirical research, Bowles argues that incentives and ethical norms often interact, and in some cases, "economic incentives may diminish ethical or other reasons for complying with social norms and contributing to the common good" (Bowles 2008, p. 1605). Interestingly, whereas incentives might be seen as increasing individual agency and autonomy (as suggested in Shahar's line of reasoning), Bowles argues that incentives may actually compromise autonomy by undermining intrinsic motivation and the sense of satisfaction associated with it (Bowles 2008, p. 1607). What's more, by framing choices in economic terms, incentives "may suggest self-interest as the appropriate behavior"

(Bowles 2008, p. 1606), shifting people's motivations away from concern for the greater good.

Bowles acknowledges that incentives aren't all bad: in certain cases, they can provide an effective "nudge" to significant behavioral change. Small taxes on bags at the grocery store provide one example: these taxes have reduced the use of paper and plastic bags significantly in places that have implemented them. One particularly striking example comes from Ireland, where a small tax led to a 94% decline in the use of plastic bags (Bowles 2008, p. 1609, citing Rosenthal 2008).

Social norms can work, too, and sometimes norms and incentives can work positively in concert with one other. In experimental economics games exploring cooperative behavior, opportunities for communication among participants, as well as punishment of non-cooperators, can help sustain cooperation over time and "may induce the self-interested to act as if they are civic-minded" (Bowles 2008, p. 1609). In contrast, in one experiment that imposed penalties on non-cooperators, people behaved more selfishly than in a parallel treatment that lacked penalties, but provided opportunities for participants to communicate (Bowles 2008, p. 1607). This suggests that communication and social connectedness may contribute positively to addressing collective action problems. The take-home message is that incentives should be designed with care, so as to avoid undermining intrinsic ethical motivation and the power of social norms.

But can't social norms generate a backlash? Shahar suggests that the answer is yes, and it seems clear that he's right: in certain cases, new norms generate resistance and backlash. The U.S. Civil Rights movement generated significant backlash, for example (Hayter 2023) and a related backlash has emerged in response to norms of diversity, equity, and inclusion emerging in American public schools, with groups such as Moms for Liberty emerging to oppose curricula that "indoctrinate" children by addressing the country's history of discrimination based on race and gender (Swenson 2023). On the other hand, incentives aren't immune to backlash either. In response to growing efforts to reduce plastic bag consumption, Mississippi passed a law that *prevents* local governments from instituting plastic bag fees or banning single-use bags altogether (see Gates 2018).

Whether through informal norms, legal regulations, or incentive plans, ethics inevitably comes into play. Although Shahar suggests that incentives can avoid moralizing climate action, deciding which incentives to develop involves values. Relying on incentives doesn't avoid ethics, it just pushes values questions back a level: i.e., what should people be encouraged and discouraged to do? Perhaps this puts less ethical pressure on individuals, who can simply decide whether to act in accordance with the incentives – but as we've seen, this may undermine intrinsic community-minded motivations.

Shahar's final worry about individual action in accordance with climate-friendly norms is that they may appear to offer a panacea for climate change, when in fact, they are only a drop in the bucket of broader social transformation. This worry is legitimate, but it applies to *many* aspects of climate action, not just individual action or individual emissions reductions. There's always the possibility that a city facing climate-related flood control issues will build a levy or sea wall and stop at that, without developing a more comprehensive plan. Or a nation may set ambitious climate mitigation targets, pat itself on the back, then take few of the steps needed to meet those targets. As Elke Weber explains:

> Decision makers are very likely to take one action to reduce a risk that they worry about, but are much less likely to take additional steps that would provide incremental protection or risk reduction. The single action taken is not necessarily the most effective one, nor is it the same for different decision makers. However, regardless of which single action is taken first, decision makers have a tendency not to take any further action, presumably because the first action suffices in reducing the feeling of worry or vulnerability. Weber found that farmers who showed concern about global warming in the early 1990s were likely to change either something in their production practice (e.g., irrigate), their pricing practice (e.g., ensure crop prices through the futures market), or lobbied for government interventions (e.g., ethanol taxes), but hardly ever engaged in more than one of those actions, even though a portfolio of protective actions might have been advisable.
>
> (Weber 2010, p. 339)

So distraction and single-action bias deserve attention, but they are not problems confined to the domain of individual action, and knowing about single-action bias may enable steps to combat it. Although it may be true that some climate activists "gravitate toward easy, conspicuous forms of activism that soothe their conscience and improve their reputation" (Shahar, p. 71, this volume), this is not the form that individual action must take. And while *certain* norms may exacerbate these tendencies, not all norms need to do so. Clearly, it doesn't make sense for climate activism to focus on promoting norms that change people's behavior in ways that make little material difference to climate change. On the other hand, norm entrepreneurs might consider the kinds of norms that can create synergies between individual behavior and systemic change. Maybe what we need are smart norms, or the right norms, not total abandonment of norm-based approaches to climate change. The norm that I have suggested – "almost everyone has reasons and opportunities to do something about climate change" – seems neither excessively prescriptive nor unusually prone to reinforce single-action bias. More generally, empirical research may help to clarify the kinds of norms that are most effective. For example, descriptive norms – citing others' environmentally friendly behaviors – may actually be more effective in motivating people than injunctive (directive/ prescriptive norms) (Hornsey and Fielding 2020). To take just one example, research has shown that in hotels, signs stating that most guests reuse their towels may be more effective in promoting towel reuse than signs encouraging guests to reuse their towels to benefit the environment (Hornsey and Fielding 2020, citing Goldstein et al. 2008). Research more generally suggests that positive social norms can be powerful motivators (Markowitz and Shariff 2012).

In developing strategies to address climate change, Shahar's discussion of norms and incentives offers some important insights: (1) it's not helpful to impose a narrow, moralizing approach to individual action that suggests everyone needs to adopt identical lifestyles; (2) to the extent possible, development of both norms and incentives should consider the risks of backlash and polarization; and (3) where climate action focuses on individual emissions reductions or the promulgation of social norms that support climate-friendly behavior, it is important to consider how these actions or norms can catalyze further action rather than stalling it. When thoughtfully constructed and paired, both norms and incentives can be effective in generating social change that can constructively address climate change. Although I agree with Shahar that individual climate-related obligations should not be defined by a set of narrow or confining lifestyle prescriptions, I don't think this shows that norms are irrelevant to positive change, or that individuals lack reasons to act to address one of the most significant challenges we currently face.

4. Conclusion

Putting together the pieces of Dan Shahar's multifaceted arguments, I believe that issues of autonomy lie at the heart of his concerns. Shahar worries about shaming or coercing people into lifestyles or vocations that don't suit their abilities and interests, and he suggests that individual climate action is therefore optional. I argue, in contrast, that many people have reasons to do *something* about climate change, but what specifically they ought to do may vary significantly, and their actions may or may not focus on individual emissions reductions. Individuals can act in the context of communities and institutions of which they are a part, so individual action need not focus primarily on personal carbon footprints. Nevertheless, individual emissions reductions can make a difference, because they can contribute to broader cultural change and have the potential to catalyze such change. What's more, individual reductions can be both a *demonstration* and *reminder* of one's commitments to climate action – they can fortify one's own commitments and have communicative value in expressing that commitment to others.

Because significant action on climate change can begin from where we are, taking advantage of our particular skills, relationships, and institutional engagements, such action need not be deeply onerous, and shame and guilt need not be the primary tools motivating oneself or others to act. There are exceptions: some actions may be beyond the pale – like intentionally spreading falsehoods about climate change (see, e.g., Oreskes and Conway 2011), or killing environmental activists (Global Witness 2022; Greenfield 2022) – and shaming and holding accountable those who are actively and intentionally causing these harms seems appropriate. More broadly, though, I think our focus should be forward-looking: how can we together build a world that enables everyone to flourish? And what can I do – given my positionality, interests, skills, and relationships – to contribute to that world? Not all of us need to be full-time climate activists, but most of us have reasons and the capacity to do something, starting from where we are.

Chapter 4

Reply to Hourdequin

Defending Specialization and Division of Moral Labor

Dan C. Shahar

Contents

1. Individual Inefficacy and its Significance   115
2. Reasons of Integrity   121
3. Reasons of Relationality   125
4. A Softer Line?   128
5. Conclusion   131

Marion Hourdequin argues in several ways for the claim that individuals should act on climate change. The first line of argument (which Hourdequin considers least persuasive) highlights the impacts we each stand to make by altering our behavior. The second links our *integrity* with our responses to this problem. The third focuses on how climate action affects our *relationships* with others. All three lines of argument try to show individuals have a moral responsibility to act on climate change.

In this reply, I'll seek to persuade you of two main things. First, it's possible to reconcile many of Hourdequin's claims with the position I defended earlier, according to which individuals *aren't* morally obligated to participate in climate action. Although this book is framed as a debate in which Hourdequin is my adversary – and although we do disagree about essential points – I think much of what she says is correct. By taking her insights onboard, my arguments can be made stronger than they were in my initial presentation.

On the other hand, to the extent Hourdequin and I disagree, I also think the position I developed in my opening statement can withstand her competing claims. As I'll try to show, Hourdequin identifies vital considerations that *can* support climate action, but

DOI: 10.4324/9781003146438-6

she interprets them too narrowly by casting them as requiring us to respond to this issue specifically. As I argued earlier, climate change is only one of many problems we can devote ourselves to tackling. Although I believe we should applaud those who participate in climate action, we also should acknowledge that other specializations are compatible with being an ethical person.

I. Individual Inefficacy and its Significance

Let's begin by examining Hourdequin's first line of argument, which highlights individuals' ability to mitigate climate change by altering their behavior. Hourdequin presents this argument in opposition to a simplistic consequentialist outlook[1] according to which: (a) individuals' responsibility to act on climate change hinges solely on their ability to influence its impacts; and (b) no individual can mitigate climate change substantially through unilateral action. Because this outlook treats climatic impacts as the sole measure of moral obligation, it follows that if no one can influence these impacts substantially, then no one has a responsibility to change how they behave. In fact, if we grant that people typically *benefit* from contributing to climate change, this outlook might even imply it's *misguided* for individuals to stop doing so since this would sacrifice their wellbeing without corresponding gain.

Hourdequin rightly dismisses this analysis of climate ethics. For one thing, as she demonstrates, there are many reasons to doubt morality always boils down to toting up costs and benefits. For another thing, climate action often produces significant effects that go beyond its direct influence on climatic outcomes. (More on these points later.) However, as a purely factual matter, it's also questionable whether individuals' contributions to climate change are as directly inefficacious as this argument suggests. For instance, John Nolt claims the average American's lifetime greenhouse gas emissions cause the suffering or deaths of one to two future people (Nolt 2011). Along similar lines, John Broome estimates the average citizen of a wealthy country destroys about six months of healthy human life and imposes between $19,000–65,000 in social costs (Broome 2012, pp. 74–75).[2] Hourdequin hesitates to place too much emphasis on claims like these. As I mentioned earlier, her main arguments revolve around concerns about integrity and relationality. Still, the impacts Nolt and Broome describe raise important questions for a position like mine, according to which it's okay to go on producing these effects provided we tackle other important problems.

In my opening statement, I granted that if we *individually* caused morally significant harm when we engaged in climate impacting activities, this would be a strong reason to demand responses from us. It was a crucial premise of my argument that climate impacting activities typically are benign when considered in isolation, becoming problematic only in the context of broader patterns of behavior that cause harm collectively. At least on the face of things, Nolt's and Broome's calculations seem to imply this premise was false: each of us causes morally objectionable harm on our own.

To address this concern, we need to dig into the details of the impacts Nolt and Broome identify. Although these authors talk in terms of causing suffering and death, shortening lives, and imposing costs, the truth about each person's impacts is subtler. We can begin to appreciate the subtleties by imagining someone adding a droplet of food coloring to a large swimming pool. As swimmers churn the pool water, the molecules in the droplet diffuse, making it impracticable to differentiate the food coloring molecules from the rest of the water. Yet, even if we could locate these molecules, this wouldn't tell the full story of the droplet's effects. At a microscopic level, the food coloring molecules will influence the behavior of other water molecules nearby; these molecules will influence other molecules near them; and these tiny perturbations will travel throughout the pool. Within a short time, not only will the food coloring be completely diffused, but the behavior of every molecule of pool water also will be slightly different than it would have been in the droplet's absence. Although these impacts will be tiny and imperceptible, they will also be ubiquitous.

Like in the swimming pool case, the difficulty of assessing individuals' climate impacts goes beyond tracing invisible gases in a vast global atmosphere. The deeper challenge is that the climate system is complex and chaotic, and our tiny perturbations reverberate throughout the system over time. Even if we could see perfectly into the future, we would not be able point to one to two specific victims who had their lives shortened, experienced misery, or endured damages because of *our* actions. Instead, as the influences of our emissions ripple throughout the climate system (alongside billions of others'), they'll influence the entire array of climate impacts, resulting in a distribution of burdens and benefits that differs subtly from what it would have been in our absence.

It would be impossible to trace these specific changes in practice,[3] and it's also impossible to anticipate whether they'll good or bad. To make the point vivid, picture this: in Albania in the year 2174, a storm cloud begins dropping its rain momentarily earlier. Off the coast of Bermuda in 2693, a hurricane's track shifts by a tiny fraction of a degree. In Chile in 2850, a breeze through a vineyard is imperceptibly warmer. Differences like these could have no moral significance, or they could produce serious impacts – for worse or for better. For instance, the redirected hurricane could make landfall on a more (or less) populated stretch of beach, yielding the single most significant consequence we ever produce. Yet, we can't know these fine-grained details – and the future people who endure climate change's impacts won't be able to know them either.

To the extent we can say anything concrete about the connection between individuals' emissions and specifically *negative* consequences, it's only by stepping back and considering the statistical relationship between emissions and impacts. If we grant it's generally worse to have more climate change rather than less – as we should – this implies the average effect of a person's emissions is to make the climate system slightly more dangerous.[4] As far as we know, this average increase in climatic danger attributable to a single person's emissions is extremely small. However, since greenhouse gas emissions stay in the atmosphere for a very long time, these tiny increases in risk will affect many billions of people over many hundreds of years. Even though the added threat to any individual is minuscule, a mathematical aggregation of these billions of tiny threats yields the types of figures Nolt and Broome report.

To illustrate: Nolt aggregates each emitter's impacts on the roughly 100 billion people climate change will affect over the next 40 generations (Nolt 2011, p. 8). He says we can expect to see an average of one to two additional people endure substantial burdens because of each American's lifetime emissions. Again, because of the climate system's chaotic nature, this doesn't mean there will be one or two future people who could point to each specific American as responsible for their hardships. Instead, it means that, *on average*, the 100 billion people who live over the next 40 generations will face an additional 0.000000001–0.000000002% chance of substantial harm because of what each current American does over their lifetime.

For our purposes, it doesn't matter whether these numbers are precisely correct. Nolt cautions against putting too much stock in them, insisting his primary aim is to demonstrate the likelihood of substantial impacts (Nolt 2011, p. 9).[5] In discussing Nolt's work, Hourdequin expresses reservations as well. But if we grant climate change is dangerous and it's generally worse to have more of it, there must be *some* average level of added risk we can attribute to each emitter. The pertinent question is how we should assess the moral significance of creating such tiny average risks for such large numbers of people.

Nolt's and Broome's discussions suggest it's illuminating to treat these aggregated statistical risks as analogous to ordinary harms of similar magnitudes. In other words, if we know an average person's lifetime emissions will expose 100 billion people to an additional 0.000000001–0.000000002% chance of harm, we can think about this as morally equivalent to each person harming one to two victims directly.

Yet, this equivalency is open to question. For one thing, it seems salient that if you directly caused one or two people to suffer, those victims would be justified in blaming *you* for their burdens. By contrast, for each person who experiences a climate-mediated hardship in the decades following your emissions, their chances of experiencing that burden would have been virtually identical if you had never lived. No person living now or in the future will be positioned to level a meaningful complaint against *you* for exposing *them* to climatic harm.

On the other hand, one might argue that even if no future person will be able to complain *your* emissions harmed *them*, the sheer fact one or two additional people will be harmed is enough to establish wrongdoing. To illustrate, imagine there's a giant lottery ball machine filled with 100 billion balls, each corresponding to one person in the next 40 generations. A certain number of balls is drawn at random, and each selection results in serious harm for the relevant person. If the machine is bumped, one or two additional balls will be selected. However, the jostling also may alter which specific balls are chosen. Thus, if you bump the machine, the chaotic effects on the balls' behavior may mean no specific person could aptly blame *you* for causing *them* to be harmed. Even so, the fact you triggered more balls to be selected may seem sufficient to ground a valid complaint against you for causing additional harm overall.

Crucially, such a complaint would seem viable even if we incorporated certain additional details to make the example more like climate change. For instance, suppose nearly everyone in your community bumps the lottery ball machine, and this is seen as a normal part of life. Avoiding the bumps would be inconvenient. Many who bump the machine would prefer to ignore their impacts and focus on tackling other problems. Even if we add details like these, the prospect of causing additional balls to be drawn (and hence additional people to experience serious harm) seems like a powerful reason to avoid bumping the machine.

If this illustration provided an apt metaphor for how individuals influence the climate system, it would go a considerable way toward showing we must restrict our contributions to climate change. Causing additional balls to be selected seems like a wrongful instance of harming, even if we can't neatly connect specific victims' injuries to specific bumps to the machine. However, the feature of this illustration that supports our intuition – the selection of additional balls – is an oversimplification. To speak of extra balls being drawn is to imply there *is* a specific increase in damage we can link to each person who bumps the machine (i.e., the victimization of one to two additional people). Yet, we've already seen this is not how things work with climate change. The climate system's chaotic nature is not limited to determining *who* will be harmed by our actions, as the lottery ball machine example suggests. It also determines *whether* our actions are harmful at all, as opposed to neutral or even beneficial.

Suppose we modify the illustration to reflect our relationships with climate change more faithfully. Now, bumping the lottery ball machine has the potential to change how many balls are drawn, but it's impossible to know whether a specific bump will cause the quantity to go up, down, or remain unchanged. Moreover, each person bumps the lottery ball machine not once but many times over their life. We know that, as large numbers of people bump the machine over time, the number of selected balls tends to increase. Thus, it's possible to say that, *on average*, each person's lifetime bumping correlates with the selection of an additional one to two balls. However, this relationship is purely statistical. For any given person, it's plausible *their* bumps will have no impact on how many balls are drawn, and it's also plausible they'll cause fewer balls to be selected rather than more. These uncertainties apply not only to the totality of each person's bumps but also to each individual bump.

My sense is that when we incorporate these details into the illustration, it becomes reasonable to doubt whether an analogy to ordinary harming is apt. This seems especially plausible when we assess climate-impacting actions individually instead of aggregating them over a lifetime. To give one concrete illustration: Nolt bases his calculations on the assumption that an average American emits 1,195 tons of CO_2 during their lifetime (Nolt 2011, p. 5).[6] Earlier in this book, we saw the choice to drive a gas-guzzling pickup truck instead of a fuel-efficient hybrid may cause an additional 6.5 tons of CO_2 emissions per year. This latter figure is based on assuming a driver travels 15,000 miles per year, which works out to about 40 miles per day. Each trip this hypothetical driver takes introduces tiny perturbations into the climate system, producing unknowable impacts on billions of people over hundreds of years. If the driver switches to a more fuel-efficient vehicle (or reduces their travel distance), each gallon of gasoline conserved will prevent about 20 pounds of CO_2 from entering the atmosphere (U.S. EPA 2024).[7] If we stick with Nolt's figures, each of those 20-pound increments of reduced emissions stands to mitigate future people's risks of climatic harm by something in the range of 0.000000000000008–0.000000000000016% each. And this is just an *average* figure: the true impact is fundamentally unknowable and could be bad, neutral, or even good.

To my mind, it's unhelpful to try to capture such actions' moral significance through an analogy to inflicting harm directly. Given the climate system's chaotic nature and the diffuse and unknowable character of our marginal impacts, it's unilluminating to speak as if our climate impacting activities produce objectionable harm all on their own. Hourdequin is right: those who want to make a moral case for changing individuals' behavior would be wise to seek arguments that don't revolve around our direct influences on climatic outcomes.

Before moving on, it's worth making one further point about the estimates we've been discussing. Nolt and Broome focus on the average effects of individuals' *greenhouse gas emissions* on future people. This is unobjectionable as far as it goes. However, these estimates don't reflect the full impacts of individuals' *activities* on future people. Each of us shapes the world in other ways than by polluting the atmosphere. As I noted in my opening statement, mainstream projections express broad agreement that future people will live longer, healthier, more prosperous lives on average than people today. This is because future people will inherit the fruits of our innovation and labor, and these advantages will overbalance – again, on average – the perils they face from problems like climate change. If future people will be better-off on average due to current people's activities, it follows that our average total impacts on future generations are positive, not negative. Our contributions to climate change *deduct* from these average benefits, but they don't overbalance them.

When we hear estimates like Nolt's and Broome's, it's easy to leap to the judgment that our legacy to future generations will be of having made the world a worse place to live. Certainly, dangerous climate change will be a salient part of what we bequeath our descendants. However, our global civilization has also been rapidly lifting its members out of poverty, expanding its recognition of individual rights, improving its scientific understanding, and pursuing countless other worthy goals. These efforts also are part of what future generations will inherit from us.

When we reflect on the legacy we want to leave behind, I think we'll discover we don't want future generations to wish we had never lived. Speaking personally, I wouldn't want my life's work to be captured in the lament, "We'll never know the precise effects of the greenhouse gases he produced, but it would be nice to have 1,195 fewer tons of CO_2 floating around." The position I've been defending in this book is that there are other ways to redefine one's legacy besides tackling climate change. If we pick our causes and devote ourselves to advancing them in the way I've discussed, we can leave the world a better place than it would have been without us. If we do this, I claim future generations will have reason to be glad we existed even if we don't also act to restrict our carbon footprints.

2. Reasons of Integrity

As I mentioned at the outset, Hourdequin's main arguments for climate action don't focus on single emitters' impacts. Instead, Hourdequin thinks the more persuasive case for individual responsibility to act comes from considerations linked to integrity and relationality. Let's turn now to these ideas, starting in this section with integrity.

Hourdequin draws on the work of Robert Audi and Patrick Murphy to highlight two aspects of integrity. First, a person of integrity will display a high degree of *integration* among their values, character traits, and life projects, aligning these elements to form a unified, cohesive whole. Second, a person of integrity will base their actions consistently on their commitments, making those commitments *integral* to how they live. Alongside these points, Hourdequin adds a third claim: a person of integrity also will be *intelligible* to others in the sense that their firmness of character will enable others to understand and appreciate why they do what they do. According to Hourdequin, these three characteristics differentiate the person of integrity from someone who is (and appears to others to be) a *hypocrite*.

Crucially, Hourdequin's account of integrity implies people may have strong reasons to do or avoid certain things even if no major external consequences are at stake. To illustrate the idea: an honest person doesn't need to know the specific impacts of telling a lie to know they shouldn't do it. More generally, a person of integrity will consider certain commitments central to who they are and how they live, and fidelity to these commitments often will drive their actions even when nothing else hangs in the balance.[8]

Building on this understanding, Hourdequin argues a person of integrity characteristically will act on climate change.[9] Such a person won't express hollow distress about the issue while doing nothing to address it. Nor will they treat the convoluted causal chains between their behaviors and climate impacts as excusing them from responsibility. Instead, they'll *integrate* climate concerns into their life and make them *integral* to how they conduct their affairs. This will lead naturally to climate action, whether in the form of reducing their contributions to the problem, pushing for collective solutions, or both. In turn, these activities will make the person *intelligible* to others as someone who cares genuinely about climate change.

Although I disagree with Hourdequin's contention that integrity demands climate action, I first want to say there is much in her argument I endorse. Integrity is an essential value we should prioritize even when no major external consequences are at stake. It's important to live an integrated life and to base our actions on our commitments. People who act on climate change often will manifest integrity when they do so. And if people who care about climate change don't act on the problem, there's a risk others will view them as hypocrites. On all these points, Hourdequin and I agree.

Despite granting all this, I don't think Hourdequin is right to treat climate action as a *test* of integrity in the way she does. My disagreement hinges on two main points. The first point is connected to the case for specialization and division of moral labor I presented in my opening statement. We live in a problem-filled world, and no one can respond effectively to each important issue. People will be most helpful if they focus on a limited number of problems while declining to act on others. Once someone chooses specific causes and strategies to pursue, it makes sense to expect these choices to become integrated into their life and integral to how they act. Yet, by the same token, a cause someone declines to adopt may not become integrated or integral in the same way. In these latter cases, I contend the absence of action on a specific issue won't be evidence against a person's integrity, and it may even weigh in the opposite direction.

To illustrate, imagine a hypothetical person, Darren, decides to focus his altruism on protecting drug offenders from overly punitive prison sentences. In line with Hourdequin's account, we sensibly may expect Darren to incorporate this commitment into his life in various ways. If Darren claimed dedication to this problem but never did anything to address it, this would be grounds to call him a hypocrite. Yet, such judgments seem connected with Darren's decision to focus on this issue. If Darren chose not to tackle certain other problems – say, human trafficking, pancreatic cancer, or factory farming[10] – it's much less clear his inaction on those problems would tell against his integrity. In fact, it may even be a *sign* of Darren's integrity that he doesn't try to "do something" about every problem that comes to his attention and remains focused instead on the specific issue(s) he's chosen.

This appeal to specialization gets us partway to seeing why I reject the connection Hourdequin envisions between climate action and integrity. Although integrity may bind *certain people* to act on climate change – namely, those who choose to devote themselves to tackling it – it's equally possible for people of integrity to focus on other issues instead. Whatever issue(s) a person takes up, we may expect them to integrate the resulting commitments into their life and make them integral to how they act. But it's a mistake to think every person of integrity will grant such a place of importance to climate change specifically. If someone like Darren neglects climate action because he has decided to focus on unjust sentencing instead, I see no reason to impugn his integrity on that basis.

So far, the objection I've been raising grants Hourdequin's account of integrity and argues it's possible to manifest this ideal without participating in climate action. However, my second point of contention questions whether she has characterized integrity correctly from the start. Specifically, we have seen that Hourdequin postulates a link between integrity and *intelligibility*, claiming living with integrity will enable others to understand and appreciate one's actions. Although I accept many aspects of Hourdequin's account of integrity, I believe she is mistaken on this point.

To see why, consider how an uncharitable observer, Elmira, may evaluate Darren's behavior in the case we've been discussing. Since Darren does nothing about human trafficking, Elmira may conclude he either doesn't care about this issue or lacks the integrity to translate his values into action. Elmira may draw similar inferences from Darren's inaction on pancreatic cancer, factory farming, and countless other problems, including climate change. Of course, if Elmira knew about Darren's choice to specialize in tackling unjust sentencing, she might resist such judgments (especially if she shared my outlook on the division of moral labor). But if she doesn't know these details – or doesn't appreciate the value of specialization – it shouldn't surprise us if she passes judgment on Darren anyway.

Would Elmira's assessment be evidence Darren lacks integrity? To my mind, the answer is no. If Darren is making adequate efforts to advance his chosen cause, he shouldn't be ashamed if Elmira misinterprets his focus as indifference or hypocrisy. Although Elmira's hasty judgment may be understandable in a world where most people don't tackle *any* important problems, it remains a mistake – and not an indictment of Darren.

To generalize the point: people who specialize in specific forms of altruism should expect others routinely to misinterpret their inaction on other problems as signaling indifference or hypocrisy. These misunderstandings not only are predictable, but they also are forgivable given the genuine indifference and hypocrisy that surrounds us. Yet, these judgments don't show specialists lack integrity. Instead, they show integrity and intelligibility aren't connected in the way Hourdequin claims.

Even if judgments like Elmira's are incorrect, it's worth noting they may have *practical* significance for how people like Darren should act. Darren may find it bothersome or offensive to be misjudged, and this could give him reason to change his behavior. Likewise, Darren may stand to gain benefits or advantages (or avoid costs or disadvantages) by sending specific signals to others. Although Darren would be wise to account for practical considerations like these, I deny they are reasons of integrity. In fact, it may even be a *sign* of Darren's integrity if he decides to stick with his chosen projects despite the inconvenience of others' misjudgments.

These reflections position us to see why I'm not persuaded of the tight connection Hourdequin draws between integrity and climate action. I agree with much of what Hourdequin says about integrity, and I grant this trait *can* manifest as she describes. But I deny climate action provides the *test* of integrity Hourdequin envisions. For people who choose to focus on problems other than climate change, inaction on this specific issue provides little basis for diagnosing a lack of integrity. Although some people may find such decisions unintelligible and form negative judgments on this basis, these misunderstandings also imply no lack of integrity in their targets. Specialization is compatible with integrity, even if it's likely to be misconstrued.

3.   Reasons of Relationality

Hourdequin's other main line of argument focuses on relationality. This discussion takes as its foil an oversimplified model of rational choice according to which narrow self-interest drives everything people do. According to this model, people predictably seek to benefit at others' expense and neglect pro-social behavior when they can get away with it. If we apply this model to climate change, we can explain the problem clearly and simply. Each person enjoys the benefits of climate-impacting activities while the costs are dispersed widely. Although it might be better for all if *everyone* restrained their emissions, no individual chooses what *everyone* will do: they only choose what *they* will do. If each emitter predictably bases their choice on self-interest, they'll go on emitting until something makes it worthwhile to stop. Hence, this analysis suggests, resolving climate change can't be achieved by browbeating people into altruistic climate action: it requires fixing the bad incentives that give rise to the problem in the first place.[11]

Hourdequin sensibly complains that the apparent elegance of this "tragedy of the commons" analysis is achieved only by ignoring much about how human sociality works. Actual social life isn't just about advancing narrow self-interest. Real people appreciate the importance of "playing fair" and doing their part to sustain cooperative arrangements. They care about setting positive examples for others. They're sensitive to their communities' norms. They work to be good neighbors and citizens. There may be limits to how much people will sacrifice for others' sake, but these limits aren't fixed at *zero* like the oversimplified rational choice model suggests. Complicating details like these explain why, in countless examples from around the world, we *don't* find people ransacking shared resources until they're coerced into stopping. Instead, we find them actively formulating self-governance arrangements to prevent their communities from descending into tragedy (Ostrom 1990; National Research Council 2002).

This broader lens on human behavior helps highlight the meaningful role individuals can play in helping to tackle climate change. Those who participate in climate action aren't just lashing out against "the remorseless working of things" (Albert Whitehead, quoted in Hardin 1968, p. 1244), as the tragedy of the commons analysis suggests. Instead, they're pursuing a valid strategy for helping to address this problem. Although no individual's altruism will alleviate climate change on its own, Hourdequin rightly cautions against expecting a single monolithic "solution" (see also Ostrom, Janssen, and Anderies 2007; Lofthouse and Herzberg 2023). A successful response to climate change will be polycentric, involving diverse contributions at many different levels (Ostrom 2009a, 2012).

I have no objections to this general line of reasoning. I agree human motivation goes far beyond the narrowly self-serving caricature some authors describe.[12] Real human sociality includes numerous mechanisms for solving problems voluntarily and collaboratively instead of relying on external coercion. Ethical people can and should take advantage of these mechanisms to tackle social problems, and they should be prepared to cooperate fairly with others to sustain solutions that emerge.

To the extent I have reservations about Hourdequin's treatment of these issues, they revolve around two points of clarification. First, although I agree we can participate fruitfully in tackling climate change through the efforts Hourdequin describes, I deny we're *required* to do so. As I've been arguing, these specific forms of altruism are just some of the many ways we can exhibit the sociality Hourdequin celebrates, and it's okay to focus on other pathways instead.

Second, it's likely that at least some cooperative tools we use to solve problems other than climate change will have limited potential in this domain. For instance, in my opening statement, I expressed skepticism about the wisdom of tackling climate change by cultivating social norms. My skepticism wasn't based on rejecting this strategy in general. Instead, my point was that *in this specific context*, norm-based strategies risk being inefficient, distracting, and polarizing, such that climate activists would be wise to prioritize other approaches. We may grant norms' value for tackling other problems (as in Hourdequin's example of landscaping water use) while doubting their potential in this domain.

These proposed clarifications to Hourdequin's arguments may seem relatively minor. However, they reflect significant differences between our visions of sociality in a problem-filled world. Hourdequin claims ethical people will manifest relationality by acting on climate change. Yet, it would seem possible to enlist relational considerations like the ones she describes to support action on countless other problems as well. If people respond to all these considerations as Hourdequin envisions, the result will be the generalist approach to altruism I've criticized, where people make many small efforts to tackle many different problems in the hope that these actions culminate in large-scale changes.

In contrast, I've argued it's permissible – and perhaps even preferable – for people to specialize in tackling a narrow set of problems rather than spreading themselves thinly across many worthy causes. Individuals who do this can dramatically increase their effectiveness at advancing their chosen causes. Although each specialist leaves countless problems unaddressed, the collective results of specialization and division of moral labor can greatly exceed what generalists can achieve through unfocused efforts.

The social landscape that results from a division of labor among altruists is different from one where each person tackles numerous problems. We're familiar with this difference from economic life, where we regularly encounter people who occupy narrow roles such as farmers, doctors, teachers, etc. For the most part, doctors don't grow food; teachers don't heal people; and farmers don't educate children. In this respect, each specialist leaves countless social functions unfulfilled. Yet, when we step back and appreciate how specialization and division of labor support better social functioning overall, this preempts the temptation to characterize specialists as deficient. In complex economies, our standards for assessing social relationships make room for dramatic differences in how individuals contribute to their communities.

I've argued for extending this vision into the arena of altruistic action. Hourdequin is correct that people promote key relational values when they participate in climate action. However, I claim that in this respect, these people are like farmers, doctors, and teachers who each take specific roles in a broader division of labor. Altruists who tackle other causes carve out different social relationships, but these differences do not imply any deficiency. Just as we don't call farmers, doctors, or teachers deficient for neglecting each other's jobs, we shouldn't assume a person's choice to focus on problems other than climate change belies insensitivity to reasons of relationality.

4.   A Softer Line?

I've spent the last three sections objecting to Hourdequin's arguments. However, a charitable reader of her opening statement may observe there's a way to interpret her position that avoids many of my concerns. According to this interpretation, although Hourdequin's case may not vindicate a broad moral responsibility to act on climate change, it does establish two adjacent points: (1) everyone has *some reasons* to act on climate change; and (2) *certain people* are bound to act on climate change because of their distinctive personal commitments regarding this issue. In my view, these two points are correct. However, I also believe that if we interpret Hourdequin as advancing nothing more than these claims, her position softens into a version of mine.

Let me elaborate, beginning with the idea that everyone has reasons to act on climate change. Throughout her opening statement, Hourdequin characterizes her arguments as identifying *reasons* to participate in climate action. For instance, Hourdequin's discussion of integrity refers not to *duties* of integrity but rather to *reasons* of integrity. The same is true of her discussion of relationality. Thus far, I've presumed Hourdequin thinks these reasons will be *decisive* (in the sense of not being overridden or undercut by competing considerations), at least for typical citizens of affluent nations. However, Hourdequin's comments in her opening statement don't *commit her* to such a claim, and it's possible for her to deny it.

Saying someone *has reasons* to do something is very different from saying they *must* do it. In fact, there are things we have reasons to do but ought nevertheless to avoid. For example, suppose a nearby restaurant offers delicious food with excellent service at affordable prices. These facts give you reasons to eat at this restaurant. Even so, it doesn't follow you *must* eat at the restaurant, and it could be that you *ought to avoid* eating there. Perhaps your spouse has spent all day preparing an elaborate meal, in which case you had better go home to eat it. The delicious food, excellent service, and affordable prices still would count as *reasons* to eat at the restaurant; they just wouldn't be *decisive* reasons given your specific situation.

The phenomenon of having non-decisive reasons is common throughout our lives. If a firm offers excellent wages and benefits, these facts give you reasons to apply for a position there. Likewise, if a person is friendly, attractive, and romantically interested in you, these facts give you reasons to invite them out on a date. However, it doesn't follow you *must* apply to the job posting or solicit the date, and it could be these actions are strongly inadvisable because of other considerations. For instance, the well-paying company could be inconveniently far away. The amorous person could be your best friend's sibling. In cases like these, you might be wise to avoid certain actions despite having reasons to take them.

Reasons can fail to be decisive in multiple ways. In the examples we've just seen, the relevant reasons were non-decisive because of their *strengths*: in each case, the reasons were defeated by stronger competing considerations. However, reasons also can fail to be decisive because the course of action they support is only one among a range of options (Portmore 2012). To illustrate, suppose you stand to lose your job unless you mail a package by the end of the day. In this case, you have a strong reason to walk the package to a postal drop box down the street from your office. By the same token, you also have a strong reason to bike the package to the post office on your way home. To justify declining to walk to the drop box, you needn't invoke a consideration that's equivalently strong to your reason to keep your job. If it would be even slightly more convenient to post the package by bike than on foot, that would be enough to make the case for that option. When multiple alternatives are available, reasons can fail to be decisive in supporting a specific course of action, not because they're weak, but rather because there are other ways to respond aptly to the reasons one has.

Returning to the subject at hand, I've been disputing the existence of decisive reasons to participate in climate action.[13] Yet, this is not because it's impossible to marshal powerful reasons for participating. As Hourdequin helps us see, many such reasons exist. These reasons fail to be decisive, not because they're weak or nonexistent, but rather because they support a course of action that's only one among a range of options. I've argued that if we specialize in tackling problems other than climate change, and if we devote ourselves fully enough to our chosen causes, this can count as responding aptly to the reasons we have. This is compatible with granting we have reasons to act on climate change – even powerful ones – provided we also acknowledge these reasons often aren't decisive.

This conclusion provides a natural segue to the second claim I mentioned above, namely that for at least some individuals, the reasons for acting on climate change *will be* decisive because of specific commitments these people have. At points in her discussion, Hourdequin focuses her comments on individuals who adopt specific orientations toward the climate change issue. For example, when exploring a link between integrity and climate action, Hourdequin says her arguments "apply most directly to those people who already care about climate change and who are committed to doing *something* about it" (p. 19, this volume). Elsewhere, she highlights people who "have large carbon footprints as well as the capacity to reduce them without significant sacrifice" (p. 5, this volume). If we continue adding descriptors like these, we presumably can get to a point where someone who refused to act on climate change would be plainly misguided. For instance, imagine a hypothetical person, Felix, who cares about climate change, commits to doing something about it, and has a large, easily reducible carbon footprint. Additionally, Felix finds climate action exhilarating, has longstanding relationships with environmental organizations, knows numerous elected officials and regulators with "green" track records, graduated from college with a double-major in atmospheric science and public administration, lives in a house already equipped with solar panels, and loves cycling to work. With all these factors pointing toward climate action – including Felix's explicit intentions – it would seem foolish to dispute the wisdom of selecting this cause. Although Felix *could* brush these considerations aside and tackle other issues instead, this would seem imprudent and even irrational.

I don't dispute that, for some people, the reasons for tackling climate change will be decisive despite all I have said. Along similar lines, I can't deny some people will have decisive reasons to tackle other causes as well. Certain people will be especially well positioned to prevent opioid deaths; others will fit especially well into campaigns against child abuse; and so on. In some cases, the relevant individuals' reasons may rise to the level of *demanding* they act in specific ways. I don't mean to rule out such possibilities when I deny we are obligated to act on climate change. Instead, I intend to be speaking generally in support of diverse forms of specialization among altruists. Individual counterexamples aren't a problem for this position: they're the exceptions who prove the rule. Since most people *aren't* overwhelmingly suited to tackle climate change over all other issues, they face the choices I've described between competing forms of altruism in our problem-filled world.

5. Conclusion

As a self-described "debate" book, this volume naturally presents Hourdequin as my intellectual adversary. However, as I hope this response makes clear, there's much about Hourdequin's opening statement I endorse. Given the serious problems climate change presents, I agree we have strong reasons to consider acting on this issue. Although I don't think these reasons are decisive for all of us, I think we owe it to ourselves to reflect on them seriously. As I said in my opening statement, the fact you're reading a book about climate ethics suggests *you* may be well served to choose this specific cause. Even if people may legitimately focus on causes other than climate change, it's also legitimate – and even praiseworthy – to select climate change as *your* cause.

When it comes to those who choose thoughtfully to participate in climate action, I believe much of what Hourdequin says is correct. To the extent Hourdequin and I disagree, it's not about whether climate action is a valuable form of altruism. Rather, it's about the specificity of the considerations that weigh in its favor. In my view, numerous forms of altruism allow people to realize the values of integrity and relationality Hourdequin describes. Thus, I believe it's possible to live ethically while declining to act on climate change, not because there's nothing to be said for doing so, but rather because it's legitimate to focus on other causes instead. If we identify a set of problems to tackle, and if we devote ourselves fully enough to the causes we choose, I claim that can be enough. The fact we don't participate specifically in climate action needn't be a basis for guilt.

On the other hand, my position still requires people to do *something* to help tackle the world's many problems. Hence, Hourdequin's analysis presents a challenge to each of us. If *we* aren't doing anything about climate change, Hourdequin's arguments raise the question of whether we have integrity. Do our actions align with our commitments? Are the things we claim to care about really at the center of our lives? Hourdequin's arguments also give us reason to wonder whether we're good community members. Are we doing our parts to address the problems we face? Are we setting a good example for others? Or are we selfishly prioritizing our narrow interests over the broader good?

I've argued it's possible to answer these questions satisfactorily even if one doesn't act on climate change. But unless *you* can explain what you're doing about problems other than climate change, you'll receive no defense from me. If you're like most people who take little action on any important problem, my message is simple. Either start doing the kinds of things Hourdequin advocates or figure out what you'd rather do instead.

Closing Statements

Chapter 5

Closing Statement

Individual Actions Matter for Climate Change

Marion Hourdequin

Contents

1. Introduction   135
2. Different Lenses for Viewing Climate Ethics   137
   2.1. Individual Inefficacy Revisited   138
   2.2. Isolationism, Consequentialism, and Cumulative Effects   139
   2.3. Social Reasoning, Relational Reasons, and Integrity   144
3. How Individual Actions Matter   145

1. Introduction

In his reply, Dan Shahar helpfully highlights places where our views converge, as well as where they depart from one another. I have tried to do the same. A key goal of our debate is to illuminate more clearly what's at stake in individual choices to act – or not act – on climate change, and to model constructive dialogue on these questions. As a reader, you may find one view more convincing than the other, but we hope you find important insights in each.

It is worth noting that neither view supports complacency.

Shahar argues that *everyone should do something to make the world a better place.* Choose an issue you're passionate about and pursue it. You might be passionate about supporting literacy for kids in local schools, enabling equitable access to health care in your community, improving labor conditions for workers worldwide, alleviating poverty, or protecting biodiversity. What matters is that you find *something* to pursue. Climate action is a viable and worthy option, argues Shahar, but it is not the only one: individual

DOI: 10.4324/9781003146438-8

climate action is optional, even for those who care about climate change and have resources to do something about it.

I take a slightly stronger position, suggesting that those of us who are concerned about climate change have reasons to do something about it insofar as we are able (though the shape of that something can differ significantly from person to person), and that most people have reasons to be concerned about climate change. Together, these premises suggest that *most people have reasons to do something about climate change, insofar as they are able.*

Shahar and I agree that reasons to act on climate change can be outweighed by other reasons for action. If I need to find a job to pay next month's rent, my reasons to search for a job may strongly outweigh my reasons to do much about climate change, at least for the time being. Clearly, climate-related reasons aren't the *only* reasons for action. Ethics should be sensitive to context, and climate-related reasons for action can be overridden by other reasons, as Shahar notes. That's why I tend not to speak in terms of universal "oughts" and "shoulds." Not only do I feel uncomfortable offering blanket prescriptions; I also believe that ethics involves balancing multiple considerations, and that these come together in complex ways in everyday life.

These qualifications and caveats might make it seem easy to rationalize inaction and wriggle out of ethical responsibilities. However, if we approach ethics in good faith and support one another in acting well, we can diminish this risk. Thus, although it may seem a bit weak to say that people who care about and have the capacity to do something about climate change have *reasons* to act (rather than that they *should*, or *must* act), taking this claim seriously can change how we think and what we do. If I accept that there are reasons for me to act to address climate change, I may think twice before accepting an invitation to fly across the country to give a short talk, or explore options for connecting with far-flung colleagues that involve fewer greenhouse gas emissions. I may consider more seriously the ethical reasons for eating less meat, and I may think about how I can strengthen the sustainability commitments of the institutions of which I am a part. That doesn't mean that climate change will be the *only* consideration or even the most salient one in my transportation decisions, dietary choices, or institutional efforts, but it will be part of the mix.

There are three important points to take from this initial discussion:

1. Shahar and I agree that most people[1] have reasons to do something to make the world a better place.
2. We also agree that individual action *can* make a difference. Shahar believes that by choosing an issue or cause to pursue, individuals can make the world better. I agree and further emphasize that individual action on climate change is important in a world where climate change is a pervasive, multigenerational challenge with significant implications not only for human beings, but for virtually all life on earth.
3. Like Shahar, I don't think that the only way to make a difference is through individual choices that *directly* diminish the problem in question (e.g., minimizing waste through recycling, or reducing one's personal greenhouse gas emissions). Political organizing, engagement with NGOs, efforts to raise awareness, and campaigns for institutional change can all have significant impacts. Sometimes even a single well-crafted intervention to catalyze changes in policies, institutions, or infrastructure can make a difference: a friend of mine wrote her city council about a dangerous intersection for pedestrians and bicycles, and the city took action to make safety improvements.[2] Similarly, even a single well-researched article in a campus newspaper has the potential to catalyze significant change. Don't underestimate the power of small steps.

2. Different Lenses for Viewing Climate Ethics

Although Shahar and I agree on many points, we approach ethical questions a bit differently. As I read it, Shahar focuses primarily on channeling individual interests and values toward socially positive consequences. He suggests that individuals have responsibilities to make the world a better place, but they should have the freedom to choose *how* to do so. His view combines a consequentialist ethical orientation with an emphasis on the value of individual autonomy.

Although I, too, believe that consequences are ethically important, my view emphasizes more strongly the relational dimensions of ethical life. I see individual actions as embedded in broader webs of social relations, and as such, I think it can be helpful to reframe conversations around individual action from a relational point of view. This diminishes the distinction between individual action and collective action, and it challenges the idea that the ethical implications of an individual's emissions can be understood in isolation from others (for discussion of related issues, see Hourdequin 2010; Brownstein, Kelly, and Madva 2021). From this perspective, ethics involves working together to enable mutual flourishing.

2.1. Individual Inefficacy Revisited

In his reply, Shahar argues that individual emissions – by themselves – do not contribute significantly to climate change or to associated harms. As such, he argues, there is no harm-based reason for individuals to curtail their personal greenhouse gas emissions. As Shahar (p. 116, this volume) puts it, "climate impacting activities typically are benign when considered in isolation," and only cause harm when combined with the emissions of many others. Thus, if individuals can be said to cause climate-related harm through their personal emissions, it is only in a statistical sense: "the average effect of a person's emissions is to make the climate system slightly more dangerous" than it otherwise would have been. Driving a gas-guzzling SUV for fun (see Sinnott-Armstrong 2005) or taking frequent airplane flights adds greenhouse gases to the atmosphere, but it's impossible to say how a *particular* car trip or plane flight changes the climate, both because the effects are impossible to trace and because the relatively small quantity of greenhouse gases associated with a particular trip are unlikely to result in significant climatic transformation (but see Hiller 2011, p. 352, who argues that individual car trips cause a "positive amount of expected harm," even if the quantity of such harm is modest). This is true even though all together, the millions of airline flights and the billions of car trips each year *do* contribute measurably to climate change. Shahar concludes that any given trip may slightly increase the likelihood of negative climatic consequences overall, but no particular trip causes specific climate-related harms. Thus, that "those who want to make a moral case for changing individuals' behavior would be wise to seek arguments that don't involve around [each individual's] direct influences on climatic outcomes" (p. 120, this volume).

Shahar and I agree on this conclusion, but we draw different lessons from it. Shahar suggests that the inefficacy of individual emissions means that people shouldn't worry much about their carbon footprints. If no individual greenhouse gas-emitting action causes specific climate-related harms, it seems that there is no climate-related, harm-based reason for a person to forego a particular trip, flight, steak dinner, or other activity associated with individual-level greenhouse gas emissions. Plus, Shahar argues, there are lots of ways that individuals are making the world *better* for future generations through their actions (e.g., by developing new medical knowledge and applications, new technology, building infrastructure that future generations will enjoy, sustaining vibrant economies, etc.), and "these advantages will overbalance ... on average ... the perils [future people] face from problems like climate change" (p. 121, this volume). We should get credit for the good things we do for future generations, not only for the negative impacts of global climate change.

Although I agree that focusing on the direct climatic impacts of individual emissions reductions can be unhelpful, I don't think this is because individual actions are inefficacious in a broad sense. What's more, emphasizing "individual inefficacy" may suggest that it really doesn't matter whether one emits profligately or parsimoniously. Additionally, the idea of "individual inefficacy" in relation to personal emissions may be misinterpreted to suggest that individuals can't do *anything* useful to combat climate change. But as I argue below, individual choices with respect to greenhouse gas emissions *do* matter, and individuals also can act efficaciously on climate change in many constructive ways (not only by reducing their own emissions). In thinking about climate change and individual action, it can also be helpful to focus on multiple dimensions of moral life, including how our choices shape relationships and communities, and how they matter for individual agency and integrity.

2.2. Isolationism, Consequentialism, and Cumulative Effects

Arguments for individual inefficacy have a point: it is true that one extra car trip by a single person won't change the fundamental trajectory of climate change. But is this the best way to think about the problem? Focusing on the (small or statistical) consequences of a particular car trip or plane flight is limiting both from a broad consequentialist perspective and from a more relational one. Although an individual car trip may "be benign *when considered in isolation*," ethics is about how to live *together*: it's not only about what

I should do but what *we* (both individually and collectively) should do. Consider the implications of everyone reasoning individually that their plane flight (or car trip or steak dinner) makes no difference to climate change. Collectively, this is likely to lead to a *de facto* policy of individually unlimited emissions, even though we know that taken collectively, excessive individual emissions contribute substantially to climate change.

To avoid this outcome, it is helpful to consider the effects of others reasoning similarly. Imagine that I had enough money to vacation in a distant part of the world four times a year. Should I do so? This would involve eight long-haul plane flights (four round trips), generating enough greenhouse gases to substantially increase my individual annual greenhouse gas emissions. But as we have seen, this increment of greenhouse gas emissions, even if a significant part of my personal carbon footprint, may not – by itself – appreciably change the climate or directly cause significant climate-related harm. I might therefore say to myself, "Since my far-flung vacations cause no harm, there's no reason not to take them." However, if *every* wealthy person reasons this way about their vacations, second and third homes, private jets, and luxury consumption, then there will be *lots* of unnecessary greenhouse gas emissions happening every year, and these luxury emissions *do* contribute substantially to climate change.

Rather than think of my actions in isolation from the actions of others, I can consider my actions as a member of a community of people who are each deciding how to act, and as a *type* of action that contributes to climate change. From this perspective, my flights generate luxury emissions, a type of emissions that arguably can and should be radically reduced. Seeing that my proposed vacation adds to a class of emissions that is unnecessary and needs to be reduced gives me at least a *prima facie* reason to curtail my long-distance trips.

Of course, a wealthy person might reply that all their wealthy friends take similar trips, so why should they sacrifice their vacation when others do not? The answer to this question is complex, but from an *ethical* perspective, it still seems clear that there is a reason not to take regular vacations of this kind, and there are surely alternatives. Perhaps a local vacation could be similarly enjoyable, relaxing, or edifying, for example. Or perhaps the *number* of long-distance vacations could be reduced.

In a 1979 article, philosopher Jonathan Harrison argues for the kind of reasoning I've described above under the moniker of *cumulative effects utilitarianism*. He suggests that his view is anticipated by utilitarian philosopher John Stuart Mill, who held that people should not do things that fall into a class of actions "which, if practiced generally, would be generally injurious" (Harrison 1979, p. 27, quoting J. S. Mill's *Utilitarianism*). Harrison argues that we should follow this logic in considering not only what *not* to do, but also what *to* do (Harrison 1979, p. 27). As Harrison (1979, p. 29) puts it, "What makes it our duty to act in [a certain way] is that good would result if everybody acted in this way." What I think is most useful to take from Harrison is not a strict account of duties, but a *way of reasoning* that considers how individual actions come together with those of others, and how socially oriented, cooperative reasoning can contribute to the resolution of collective action problems.

The benefit of viewing individual actions in relation to broader cumulative effects – and the dangers of considering actions in isolation from one another – can be seen by considering how isolationist reasoning can be used to break down many actions into relatively harmless sub-actions: adding one drop of oil to a small pond shouldn't appreciably damage the fish or aquatic creatures living there,[3] so adding one drop of oil is relatively "benign when considered in isolation" (to quote Shahar). But what if I add two gallons of oil to a small pond, one drop at a time? Now the pond is significantly more polluted, perhaps harming the fish and amphibians that live there. But if it wasn't problematic for me to add any of the individual drops – because no single drop made the pond appreciably more polluted than it was before that drop was added – then it seems that I am not blameworthy for adding two gallons of oil to the pond, as long as I've added it drop by drop. And yet, many people would say that I have acted wrongly regardless of whether I partition my additions into drops: whether I dumped in two gallons all at once or added it in a rapid series of drops, I've still polluted the pond, and (in the absence of compelling countervailing reasons) one ought not to pollute ponds. Similarly, if 256 people each add an ounce of oil to the pond on the same day, each person is only responsible for a small amount of pollution – but collectively, they've damaged the pond. It would be much better if none of them or only a few of them dumped oil in the pond.

Many environmental problems have this shape, where individual activities add up over time or accumulate through contributions of multiple actors. For this reason, it is critical to pay attention to cumulative effects, to the contexts in which we're acting, to how our actions now may connect with our actions in the future, and to how an individual's actions work in concert with those of others. Context matters: even in the case above, there are details that could alter whether my pollution should count as harmful or morally blameworthy (the importance of context in relation to individual climate action is also noted by Hiller 2011; Raterman 2012; Schwenkenbecher 2014). For example, if a total of two gallons of oil are added to the pond in tiny increments over a very long period of time – a century, say – the pond is unlikely to be harmed. So, if I'm the only one adding oil to the pond, and I add just one drop a year over my entire lifetime, my actions may not be morally problematic from the perspective of harm caused. On the other hand, if *everyone* adds a drop of oil to the pond each year, and this pollutes the pond, my additions may be morally problematic, because they are part of the type of action that is polluting the pond.

With Harrison, I believe that consequentialists should consider cumulative effects, and that the right question to ask in the case of one's personal emissions is not about the climatic effects of one particular individual action (which, as Shahar points out, is almost always a drop in the bucket), but about the effects of many people acting in the way one is considering. Even though I acknowledge that there are generally multiple considerations at stake, what Harrison calls cumulative effects utilitarianism can provide reasons to curtail one's individual emissions, particularly luxury emissions that add little to a person's quality of life.

In today's context, when there is a pressing need to reduce greenhouse gas concentrations in the atmosphere, it doesn't make much sense to consider excessive individual emissions in isolation. We might instead use the analogy of a very full bathtub, on the verge of overflowing and causing significant damage. Adding a cup of water to a very full (or already overflowing) bathtub not only has different consequences than adding a cup of water to an empty tub, it also has different ethical significance.[4] If we think not only in terms of our own individual impacts, but more broadly about *what each person should do* in this situation, the obvious answer is that each person should add as little water as possible to the tub, and to refrain from adding water if they reasonably can.

What might Shahar say to this line of argument? He would likely agree that the ideal response in each case would be cooperation, so that we collectively add little to no oil to the lake, stop pouring water into the overfull tub, and reduce emissions by individuals and institutions to address climate change. In a cooperative environment, individual reductions in pollution (whether oil in a lake or greenhouse gases) are likely to combine with those of others, and may inspire others to act similarly, so in these situations, it makes sense to exercise individual restraint. However, in his book *Why It's OK to Eat Meat*, Shahar (2022, ch. 5) suggests that if there is good reason to believe that others *won't* cooperate (i.e., where most people won't do their individual part to solve a problem), it may be better to focus one's attention elsewhere, especially if individual restraint is personally costly, because such restraint is unlikely to act synergistically with that of others to produce a positive result.

This is an issue on which we disagree. Shahar thinks that individual efforts to act on climate change – particular individual emissions reductions – often involve fruitless sacrifice, and that abstemious "climate-friendly lifestyles" may even dissuade others from taking up the climate cause (Shahar, conclusion of this volume). I am more inclined to think that climate-friendly actions can be an important way of affirming and expressing concern for building a livable world, and that these actions can contribute to broader social change. But I concede that context matters, and a holier-than-thou approach to individual action could certainly turn others away from the cause. Nevertheless, changes in action can lead to changes in culture and changes in institutions, and we need both to effectively respond to climate change.

In sum, focusing on individual actions in isolation can lead people to conclude that their individual emissions are relatively inconsequential, and thereby to conclude that it doesn't matter whether one emits profligately or not. Although it is true that *when considered in isolation*, an individual's emissions – particularly over a short period of time – may not appreciably alter the climate system for the worse, our individual emissions are not isolated from those of others. It is my emissions *together with those of many other agents* (from individuals to institutions) that generate climate change. Thus, we need to consider not only our own actions, but how these actions fit together with those of others to exacerbate or diminish climate change. By isolating individual actions and individual reasoning from those of others, arguments for individual inefficacy miss the big picture and the value of "we"-oriented forms of reasoning.[5] These forms of reasoning resonate with many relationally and socially grounded ethical models.

2.3. Social Reasoning, Relational Reasons, and Integrity

Although Jonathan Harrison is a utilitarian and thereby also a consequentialist, the form of reasoning he recommends shifts from an individual isolationalist model toward a more social and relational one. By acknowledging that whether I ought to do something is not a matter only of whether that action alone has negative consequences, but of whether everyone acting in this way would be harmful, Harrison recommends a social form of reasoning, asking *what each of us should do as individuals who are part of broader collectives.*

This form of reasoning is particularly helpful in relation to individual emissions, because individual emissions can seem inconsequential, even though individual emissions choices matter a lot in the aggregate. Together, our individual choices to reduce emissions can mitigate climate impacts, just as increasing total individual emissions globally can exacerbate them. Notice, too, that the individual inefficacy argument applies at multiple scales, and thus the risks associated with it may extend beyond individuals. For example, an individual company or institution might reason that *its* emissions, considered in isolation, won't really affect the trajectory of climate change, so it shouldn't worry about its emissions, particularly if its actions are otherwise socially beneficial. Colleges and universities might use this reasoning to argue that they lack obligations to curtail their own emissions or to transition to sustainable energy. On the other hand, thinking about what colleges and universities *generally* should do about climate change, and what individual institutions should do to be part of that collective shift, can catalyze constructive change.

Thus far, this section has focused primarily on individual emissions. However, there are many other important forms of individual climate action, which matter in multiple ways and at multiple scales.

3. How Individual Actions Matter

From a scoped-out perspective on global problems, most individual actions – whether reducing emissions or lobbying for a policy change or cleaning up a local stream – can seem small and inconsequential. In the face of challenging problems, it can be tempting to give up, thinking that one person can't make a difference. However, this perspective is limiting. As I argue below, viewing one's actions as inefficacious is not good for individuals or for communities, and from a relational perspective, it can negatively affect the character and quality of human relations with other people, other living beings, and the broader environments in which we live.

Let's take an example to illustrate. Imagine if every teacher said to themselves, "By helping this particular student, today, I'm not really making any difference in terms of the level of education and learning in the world more broadly, so my individual efforts don't matter." By this same logic, I might reason that by being rude to a store clerk, I add very little to the overall unhappiness in the world, so it must be okay.

Of course, many of us are grateful for individual teachers who made a difference in our lives, and we know how a warm smile can make our days go better, just as a mean-spirited comment can ruin a perfectly good day. A single act of kindness or generosity may not transform the whole world, but that doesn't mean it doesn't *matter*. Small actions do matter, and we see this regularly in our everyday lives.

The tricky thing about climate change is that some efforts to address it don't have obvious, immediate, positive effects, and more generally, climate change requires thinking at multiple scales. Whereas the teacher may see the concrete, positive impacts of their efforts in the learning of the students in their classroom, a person who seeks to reduce their personal carbon footprint may never see comparable positive effects.

Nevertheless, our actions do matter locally, and they interact with the actions of others. Together, our actions shape our communities and the world in which we live. What do we want that world to be like? What kind of worlds can and should we build through our actions? A world in which rudeness is pervasive because each person sees their own actions as inconsequential? A world where teachers are unmotivated to help their students because their efforts are a drop in the bucket in relation to human learning more generally? Where those researching devastating diseases like malaria give up after calculating the low probability that *their* work, specifically, will generate a major breakthrough? To build better worlds, we need each other, and we need to be able to count on each other to do our little part, even if the immediate – or even the long-term consequences – of that part are modest or unclear.

The world in which we take our own actions to matter as part of the collective project of building good relations and creating a world in which all can flourish is not just better in terms of its cumulative consequences and transformative potential, it is better in terms of how we relate to one another, and better for individual persons.

We need to look at the big picture *and* the little picture. Doing something generous or kind for another person may not change the whole world, but it might change that person's day, or even their life. It can change the feel of a place, and of a community. Think about coming to a new place, such as town or school where you've never spent time before. Even small gestures of welcome in a new community can make a difference between feeling disconnected and alone, or hopeful and connected. A well-timed word of encouragement can make the difference between someone's giving up and their persisting – sometimes changing the course of their life for the better. My guess is that we can all remember moments where a teacher, coach, parent, friend, or sibling did or said something seemingly small that made a big difference to our lives.

With this perspective in mind, we can think a little differently about individual climate action. Each of our actions is part of a web of relations. Our actions don't stand alone; they connect to what we've done in the past and what we'll do in the future; to how we think about ourselves and our identities; and to how we show up and engage as members of the broader communities of which we're a part. We have relational reasons – reasons for action that emerge from our relational connections with and responsibilities to one another – to take seriously what each of us can do as members of broader communities to stem the tide of climate change and reduce its impacts on human and ecological communities and the individuals who comprise them.[6]

Taking seriously our own agency and efficacy can also be good for us personally. In some ways, reasons of integrity are the individual-level analog of reasons of relationality. Whereas relational ethical reasons are rooted in how our actions come together with those of others to support mutual flourishing, reasons of integrity have to do with internal relationships – within us – between our values and actions, and between our past, present, and future selves. Whereas relational reasons support harmonious interpersonal relationships, and reasons of integrity support greater *intra*personal harmony. Just as thinking about individual actions in isolation from the actual or potential actions of others undermines the prospects for social harmonies and the realization of collective goals and goods, acting in a way that is isolated from one's values, commitments, and past and future actions can create internal disconnects that undermine one's groundedness and sense of self. Relational reasons support interpersonal harmonies and reasons of integrity support intrapersonal ones.

Although reasons of integrity concern the way our values, actions, and commitments fit together, integrity also bears on our relationships with others. This is because integrity is associated with trust and reliability (see Palanski, Kahai, and Yammarino 2011 for discussion of the relationship between integrity and trust). If I say one thing and do another, or I act in a particular way one week and then do the opposite the next, I may not be a very reliable partner in collective action for change. Conversely, consistency between my values, commitments, and actions can enable the development of interpersonal solidarities that provide a foundation for action together with others to effect broader social change. Integrity thus can support the kinds of interpersonal relationships critical to effective collective action – and these relationships in turn, can support individual integrity, because integrity is often strengthened through relationships with others oriented toward similar values and goals. Social support can help us deepen and follow through with our own commitments, and working together with others can amplify our power to effect change.

Those of us concerned about climate change thus have at least two types of reason for climate action, as I have argued throughout this book. First, we have reasons grounded in our connections and obligations to others and to the local, global, and intergenerational communities of which we are a part (relational reasons), and second, we have intrapersonal reasons to act in alignment with our values and concerns (reasons of integrity). These reasons support intrapersonal and interpersonal harmonies that contribute constitutively to individual and collective flourishing.

But what about disagreement? Does a focus on inter- and intrapersonal harmonies allow space for tension and disagreement between people and even within ourselves, as we wrestle with complex and potentially conflicting values? My suggestion is that living well individually and collectively requires some degree of harmony, but not harmony without disagreement, difference, or tension. The idea that conflict and harmony can coexist and may in fact depend on one another is developed by philosopher David Wong, who writes:

> Let me roughly characterize the value of harmony as the value that promotes reconciliation and congruence between the individual's interests, the interests of others, and the group's common projects ... To seek to harmonize, then, is to try to render compatible and whenever possible mutually supportive the self's interests, others' interests, and the group's projects.
>
> (Wong 2011, p. 202)

With respect to climate change, we clearly need forms of harmony that acknowledge diverse needs and interests and seek reconciliation between them. Even among those of us who care deeply about addressing climate change, there will be disagreement on how best to proceed, how to address multiple – and sometimes competing – needs, and how to enable effective and just transitions away from fossil fuels.

Unmitigated climate change and the absence of coordinated efforts to address climate impacts will exacerbate disharmonies at multiple scales. We need to acknowledge that existing fossil-fuel-dependent forms of economic activity, infrastructure, and social organization may be supporting human flourishing in some ways, but undermining it in others, particularly for future generations, and we need to develop new ways of living and organizing societies that reestablish a better balance. Institutional change is critical, but individuals have an important role to play: if we recognize that we are fundamentally interconnected with others at multiple scales, we can see that our actions matter not just in isolation but in their contributions to the kind of world we wish to create and in relation to the values we ourselves hold and seek to realize in our lives.

Insofar as we care about human and ecological flourishing, we need to act on climate change, at multiple scales. Each of us has the potential to contribute, and to support others in their contributions. Together, can sustain a livable world.

Chapter 6

Closing Statement

Embracing Diversity among Altruists

Dan C. Shahar

Contents

1. The Limits of Specialization	149
2. Climate Change as a "Special Problem"	152
3. Specialization and Autonomy	155
4. Flexible Norms?	158
5. A Worthy Option	162

Hourdequin is right to say our views are not diametrically opposed. We agree that climate change is a serious problem and that it's good for people to help tackle it. To the extent our positions conflict, the disagreement boils down to two main points. First, Hourdequin thinks that – at least for those of us who live comfortable lives in affluent nations – acting on climate change is a characteristic mark of being an ethical person, whereas I've argued climate action is only one of many legitimate outlets for pro-social efforts. Second, Hourdequin favors norms that reinforce climate action with social pressure, whereas I've claimed the climate movement should use this approach only with great caution. In this closing statement, I'll defend my side of these disagreements and respond to Hourdequin's main concerns. I'll end with some hopeful takeaways from my contributions to this book.

I. The Limits of Specialization

Let's start with the idea that climate action is just one of many legitimate outlets for altruism. Underlying this claim is a more general perspective I embrace regarding altruistic specialization and division of moral labor. Given society's litany of problems, our limited

DOI: 10.4324/9781003146438-9

bandwidth for action, and the greater efficacy we can realize by specializing, I think it makes sense for each of us to focus on a small set of problems rather than tackling many issues simultaneously. In line with this outlook, I've claimed it can be legitimate for individuals to decline to take costly unilateral action on climate change and focus on other issues instead.

Hourdequin worries this position risks sliding down a slippery slope. If we grant it's desirable for people to specialize in diverse ways, why should we insist *everyone* engage in altruism? Why not embrace an even more radical form of specialization where certain people focus on pro-social action while others focus on self-interested pursuits?

To answer this question, we need to step back and ask why altruism is necessary to begin with. Intuitively, at least, the answer seems straightforward. I expect most of the people reading this book live comfortable, prosperous lives in a world full of serious problems. Although none of us can eliminate every problem by ourselves, we can help make progress through pro-social action. Given our many privileges, refusing to help would be selfish and wrong. Not only should we contribute, but we should be prepared to make sacrifices to increase our impacts. Admittedly, it's hard to say how much sacrifice it takes to discharge our duty to help. But this ambiguity doesn't undermine the basic idea that morality requires us to balance attention to our own interests with sensitivity to broader concerns.

If we concede living well involves a balance between self-interest and altruism, this helps clarify why it would be inappropriate for some people to live purely self-interested lives while others specialized in self-sacrifice. In such a division of efforts, the egoistic individuals would be guilty of selfish indifference to the problems around them. This would be objectionable even if there were other people who went above and beyond with altruism. In my view, heroic self-sacrifice by some doesn't get others "off the hook" for making sacrifices of their own.

As far as I can tell, Hourdequin shares my general outlook on the need to balance self-interest against pro-social action. For this discussion's purposes, we can also set aside any possible differences between us concerning *how much* people must sacrifice for the broader good. (Although neither of us specifies precisely how much altruism is "enough," we can provisionally accept any balance Hourdequin considers appropriate.) To the extent our positions diverge fundamentally, it's not over *whether* or *how much* people should engage in pro-social action. The major difference concerns *what people do* with the portions of their lives they devote to altruism.

Hourdequin believes everyone's portfolio of pro-social actions must include efforts on at least some common problems, of which climate change is one. In contrast, I think it's okay to specialize to the point of declining to act on these problems. Because the crux of our disagreement concerns how people direct their altruism – and not whether or how much they engage in altruism – I don't believe my position is any more vulnerable than Hourdequin's to sliding down a "slippery slope" toward the conclusion that some people may shirk altruism altogether.[1]

I'll have more to say below about whether climate change is a special problem that demands action from us all. However, before I turn to that subject, a few further clarifications are warranted. First, although I insist people should be prepared to sacrifice to help tackle the world's problems, I don't mean to treat sacrifice as the metric for evaluating altruistic efforts (see Schmidtz 2016). Many putatively pro-social undertakings are ineffective, inefficient, or counterproductive, swallowing time and resources without producing anything of comparable value (Easterly 2006; Moyo 2009; Coyne 2013). It would be a mistake to try to defend wasteful efforts by appealing to how much sacrifice they involve. In isolation, sacrifices are costs to avoid whenever possible. What justifies bearing these costs is the progress they enable us to achieve.

The reason I speak so much about sacrifice in this book is that it sets the *limit* on how far we must go to promote the broader good. As I said earlier about "no regrets" altruism, we should *always* help alleviate problems – including climate change – when we can do so without cost. It's only because many pro-social actions involve sacrifice that it makes sense to speak of a need to *balance* these actions against the other parts of our lives. Where such sacrifices are absent, the rationale for limiting our efforts is absent as well.

Let me clarify secondly that our duty to sacrifice for the broader good is linked to the circumstances we face. I take it to be factually true that our world is filled with problems and that virtually everyone reading this book can help address these problems through sacrifice. If these conditions didn't hold, the case for sacrifice would be correspondingly weak. For instance, altruistic sacrifice would seem unmotivated in a utopia. The same would be true if a group of superheroes stood ready to solve all our problems for us (to invoke a scenario Hourdequin describes).[2] It would also seem misguided to demand sacrifices from people who lacked any ability to make any positive impact. I advocate sacrifice not because I believe it's obligatory regardless of circumstances, but rather because I expect this book's audience to be comprised of the types of people who ought to be sacrificing.

Finally, let me note that the type of balancing I advocate reflects one aspect of living ethically and not the whole of morality. At least since Aristotle (around 350 B.C.E.), philosophers have distinguished between moral standards that involve strict requirements (e.g., against murdering) and those that involve finding a "middle ground" between undesirable extremes (e.g., being courageous rather than cowardly or rash) (Aristotle 2000, bk. II). Altruism typically is placed in the latter category: it's objectionable to be selfish, but morality also doesn't ask us to sacrifice everything we have.[3] Ethical people will find a "middle ground" between these extremes. Nevertheless, a virtuous balance in this domain doesn't license us to start lying, cheating, and stealing. Undertaking "enough" altruism is only part of what morality requires.

At one point in her reply to my opening statement, Hourdequin emphasizes we are "always 'on duty' from an ethical perspective," illustrating this by highlighting the absurdity of saying, "I was nice to three people this morning, so it's fine if I'm rude this afternoon, since I've done my part" (p. 104, this volume). On this, Hourdequin and I unsurprisingly agree. But I hope it's clear why this doesn't exemplify the balancing I recommend. We cannot justify mistreating others (e.g., being rude in the afternoon) by appealing to good behavior in other domains. Where I claim balancing becomes relevant – and where it can make sense to say we have "done our part" and will go no further – is with respect specifically to our duty to engage in altruism, not our broader responsibility to behave well.

2.   Climate Change as a "Special Problem"

Let's return to the question of whether climate change is a special problem that demands action from us all. In her comments, Hourdequin expresses sympathy toward the general claim people aren't required to tackle every issue. Still, she thinks I take this idea too far by applying it as broadly as I do. In her view, even though we may legitimately decline to act on *many* problems, there nevertheless are certain problems that demand universal responses. Hourdequin places climate change in this category on the grounds that it is "so wide ranging and significant in its effects" and "intertwined with and in many ways a fundamental component of all – or almost all – other problems we face" (pp. 99–100, this volume).

To assess Hourdequin's claim, we need to distinguish it from a related pragmatic point. In practice, we may find that because climate change is linked with so many other issues, actions we take to tackle non-climate-related problems often will address aspects of the climate crisis as well. To illustrate, suppose a hypothetical altruist, Ganavi, decides to help people trapped in poverty pursue economic opportunities in other countries. Perhaps Ganavi is attracted to this cause because she attributes her own prosperity to her forebearers' ability to seek better lives abroad. Although this motivation may have little direct connection to climate change, facilitating economic migration nevertheless may help promote climate adaptation. Hence, even though Ganavi hasn't set out intentionally to tackle the climate problem, her efforts arguably constitute a form of climate action. Generalizing from this example, we can appreciate that because climate change is linked to so many other issues, cases will be common where people aim to tackle other problems but help ameliorate the climate crisis as well.

When I say people have no specific obligation to tackle climate change, I don't mean to oppose Hourdequin on this point. To the extent diverse pro-social efforts often have the side effect of reducing climate change's severity, this is a result to be celebrated. Moreover, the complementarity between other causes and climate action undoubtedly counts in favor of activities that allow one to "double dip." My point is simply that when individuals choose how to direct their altruism, they are not obligated to *ensure* their efforts also combat climate change, and it can be okay if this alignment doesn't occur.

To illustrate the difference between Hourdequin's view and mine, consider her example of global hunger. As Hourdequin explains, climate change threatens food security around the world, and many of the gravest burdens will fall on populations that are already disadvantaged by global standards. Because of this, those who care about solving hunger should be concerned about climate change.

Likewise, fighting climate change can be a sensible way to advance the anti-hunger cause. On these points, Hourdequin and I agree.

However, climate action is not the only viable strategy for reducing hunger, and it may not always be the most effective approach. As Hourdequin recognizes, the principal source of hunger in practice is not a physical lack of food but rather a lack of effective economic and political means for hungry people to meet their needs.[4] Hence, it sometimes may be easier to insulate people from hunger by targeting economic and political disparities than by reducing threats to food production like climate change. There may even be cases where effective hunger-fighting actions exacerbate the climate problem (e.g., by increasing incomes and thereby expanding carbon footprints).

When the best means to eliminate hunger diverge from those that combat climate change, altruists will face difficult decisions about how to balance the competing considerations. Climate change's gravity and urgency undoubtedly should play a role in these decisions, and people certainly should avoid *creating* serious climatic harms as they pursue other causes. Yet, I deny altruists who wish to tackle global hunger always will have decisive reasons to prioritize climate action. Although hunger and climate change are intertwined, this does not mean efforts to tackle these problems will always be aligned. When they come apart, I claim the choice to focus on hunger can be legitimate.

We can see similar possibilities with Hourdequin's other examples. For the foreseeable future, climate change will influence virtually everything that happens on our planet, from e-commerce to insurance to education and beyond. In many domains, our best options for advancing non-climate-related causes will align with fighting climate change. In these cases, tackling climate change and other problems can and should go hand in hand. However, it seems undeniable there will be situations where the best solutions for other problems do not address climate change and may even work at cross purposes. In such contexts, I claim people need not always decide in the climate's favor to count as ethical.

All this aligns with my foregoing arguments for specialization and division of moral labor. But what of Hourdequin's observation that climate change is "so wide ranging and significant in its effects" and "intertwined with and in many ways a fundamental component of all – or almost all – other problems we face" (pp. 99–100, this volume)? To my mind, these seem like valid reasons to choose to focus one's pro-social efforts on climate change, but they do not underpin an *obligation* to do so. To echo an earlier point, we don't think it's everyone's responsibility to participate in farming, medicine, or teaching, even though each of these vocations is "wide ranging and significant in its effects" and "intertwined with and in many ways a fundamental component of all – or almost all – other problems we face." If society confronted a shortage of farmers, that would be a grave problem, and we would applaud people who responded by going into farming. But it would be misguided in such a situation to criticize doctors, teachers, or specialists in other areas who continued to focus on their roles. It would seem especially misguided to insist non-farming specialists should add a little agriculture to their repertoires alongside their other activities. Once we embrace the logic of specialization and division of labor, I believe we will see the types of considerations Hourdequin identifies don't imply everyone is obligated to engage in climate action. It's laudable to tackle the biggest, most important, and most pervasive problems. But the bottom line remains that it's okay for people to specialize in diverse ways.

3. Specialization and Autonomy

Near the end of her reply, Hourdequin comments on my position's relationship with the value of individual autonomy. She observes that if my view is correct, people have broad discretion over which problems to tackle and how. This expands their latitude to pursue projects that align with their interests and avoid forms of action they dislike. On the other hand, Hourdequin concedes that if we were required to contribute to specific causes as she claims, this would limit our latitude for choice. Based on these reflections, Hourdequin speculates my attraction to division of moral labor may be a product of my desire to avoid sacrificing autonomy. On the other hand, she expresses comfort at the prospect of limiting discretion to promote appropriate responses to climate change.

Hourdequin is right that I consider it an attractive feature of my position that it supports broad latitude for choice. However, I don't think she is right to see a concern for autonomy as central to my arguments, and I also don't believe my position depends on being highly motivated by this value. I take it to be a general fact about morality that our obligations limit the proper scope of discretion.

It does not count as an attack on autonomy to say we ought to respect others' rights, obey vital norms, and contribute fairly to public goods. Indeed, any ideal of autonomy worth cherishing will see fulfilling our duties as an *expression* of autonomy (i.e., living in accordance with our own will) rather than a diminution of it.[5] Thus, if morality required individuals to act against climate change, a concern for autonomy would not provide a basis to resist this conclusion. It's only because I believe morality issues no such command that I take the position I do.

It also bears noting that although specialization and division of labor expand autonomy in certain respects, they threaten it in others. To see this, consider how an economy in which people fill a variety of professional functions differs from one in which people specialize in narrow jobs. In the non-specialized economy, each person has opportunities to engage with many types of work, positioning them to develop diverse skills, experiences, understandings, and sensitivities. By contrast, although people in the specialized economy may become more efficient at their chosen disciplines, this can come at a cost to their personal growth. As Adam Smith lamented during the industrial revolution:

> In the progress of the division of labor, the employment of ... the great body of the people, comes to be confined to a few very simple operations, frequently to one or two. But the understandings of the greater part of men are necessarily formed by their ordinary employments. The man whose whole life is spent in performing a few simple operations, of which the effects are perhaps always the same, or very nearly the same, has no occasion to exert his understanding or to exercise his invention in finding out expedients for removing difficulties which never occur. He naturally loses, therefore, the habit of such exertion, and generally becomes as stupid and ignorant as it is possible for a human creature to become.
>
> (Smith [1776] 1904, V.I.III.2)

While we may bristle at Smith's characterization of eighteenth-century laborers' intellects, the danger he sees is important. Specialists must guard against becoming so pigeon-holed that they become stunted as individuals and citizens. True flourishing depends upon diverse experiences that promote development along many dimensions. This is the motivation behind Karl Marx's famous comment:

> [A]s soon as the distribution of labor comes into being, each man has a particular, exclusive sphere of activity, which is forced upon him and from which he cannot escape. He is a hunter, a fisherman, a shepherd, or a cultural critic, and must remain so if he does not want to lose his means of livelihood; while in communist society, where nobody has one exclusive sphere of activity but each can become accomplished in any branch he wishes, society regulates the general production and thus makes it possible for me to do one thing today and another tomorrow, to hunt in the morning, fish in the afternoon, rear cattle in the evening, criticize after dinner, just as I have a mind, without ever becoming hunter, fisherman, cowherd, or critic.
>
> (Marx [1846] 2000, 185)

Marx's utopian vision has proven unrealizable in practice, and societies that have tried to experiment with communism have invariably retained specialization and division of labor in their economies. Still, we can sympathize with Marx's frustration with the institution and his desire for a better alternative. The difficulty both Smith and Marx appreciated is that specialization increases efficiency so much that those who refuse to specialize will struggle to compete. Especially in the economic arena, where people must work to support themselves and their families, the imperative to specialize may be inescapable regardless of its spiritual consequences.

Pro-social actors are generally under much less pressure to specialize since they contribute at their own expense. Thus, it is unsurprising many altruists find it appealing to take a rounded approach that allows diverse forms of engagement as well as personal development across many dimensions. However, when carried too far, this approach becomes dilettantism, with people dabbling in issue areas more as a form of self-enrichment than as a coherent strategy for effecting change. If altruists are serious about making the world a better place, they must typically submit at least somewhat to the unpleasant discipline of specializing.

My call for altruistic specialization and division of moral labor thus implies a limit on discretion as well as an expansion of it. True, I claim those who specialize may engage with certain issues and activities while neglecting others altogether. However, I also claim they should *focus* and not flit among problems and strategies whenever their inclinations recommend it. Thus, from the standpoint of autonomy, this outlook both gives and takes. It licenses us to prioritize issues we care about and dismiss others as "not our problems." But it also implies that when we resolve to focus on specific activities, we must commit in a way we may come to find constricting and frustrating.

These reflections lead naturally to thoughts of finding a middle ground between narrowness and dilettantism. Although I claim it's legitimate and even desirable to specialize, this approach has personal drawbacks that make it reasonable to reflect on how far down this road one wishes to go. Adopting a disciplined focus may enable people to improve their efficacy dramatically, and to the extent the purpose of pro-social action is to make the world a better place, this should be a weighty consideration. But since excessive specialization can be stifling, we may sensibly wish to sacrifice some degree of efficacy to broaden our engagement with the world's problems. I believe conscientious individuals can navigate these tradeoffs in diverse ways, including by spending extra time and resources on less impactful actions they find more personally stimulating. Again, however, I deny people always must resolve these ambiguities in a way that includes climate action.

4. Flexible Norms?

One troubling feature of my position concerns its implications for those who wish to lean on morality to solve problems like climate change. If everyone were obligated to help tackle such problems, this would provide a resource for resolving them since anyone who failed to contribute could be criticized and pressured to change. By contrast, my position characterizes actions on problems like climate change as just some of the many ways pro-social actors may focus, such that each movement must compete with numerous others for scarce attention and resources. (As I noted in my opening statement, it is incontestable that movements face such competition *empirically*. What my position adds is that this is as it should be and not indicative of moral failure.)

If the climate movement cannot insist on altruistic contributions from all, this makes the task of tackling the climate crisis more daunting. By "more daunting," I don't mean "impossible," since we have seen there are other potential solutions that do not rely on costly unilateral action from everyone. Still, removing moral pressure from the climate movement's toolkit would eliminate one pathway many (including Hourdequin) have been keen to exploit.

As Hourdequin helps us see, backing away from moralizing can come with practical drawbacks. Individuals who see themselves as "doing the right thing" often will go to great lengths to perform actions they otherwise would prefer to avoid. Hourdequin illustrates this point well with an example of parents at a daycare she draws from Samuel Bowles (2008, p. 1605). In Bowles' example, parents who believed it was inconsiderate to leave their children past the announced pickup time tried hard to ensure they arrived promptly. Yet, when administrators made it seem parents could excuse tardiness by paying a fee, many stopped arriving on time and exchanged their money happily for increased flexibility. To create the same level of compliance through incentives that they had formerly achieved through moral pressure, the administrators might have needed to resort to extreme financial penalties or other unpalatable sanctions.

There is little doubt embracing my arguments could have similar effects in the climate arena. To illustrate, environmentalists often go to great lengths to limit their carbon footprints out of a conviction that this is the right thing to do. If they felt it was permissible to emit as much as they pleased (say, provided they contributed adequately to other causes), many might stop constraining themselves as much as they do. For instance, if a morally motivated Prius owner became convinced it could be equally ethical to drive a large, inefficient SUV, we should not be surprised if they opted for the latter even if they had to pay more for gas. Similar points apply to people who bike instead of driving, avoid air travel, or refrain from heating and cooling their homes to comfortable temperatures due to ethical concerns.

We can blunt these worries partially by noting that even without moral *commands* to reinforce climate action, there still would be a case for seeing climate efforts as ethically salient. As I've said, climate action is a valuable way to help tackle an important problem. Even if we relinquish the notion everyone is *obligated* to tackle climate change, we still can applaud those who take up this cause. Yet, there is little doubt many environmentalists act as they do from a specific sense of duty. Persuading them it can be legitimate to focus on other causes may lead them to reduce their efforts in this domain.

How big of a problem is this? To some extent, the answer depends on what else is done to control climate change. In a world with little meaningful political or technological progress on this problem, it would be lamentable if many who tackle it altruistically began abandoning their efforts. At the other extreme, this might be no problem at all in a world with a well-functioning climate policy regime. For instance, under an effective cap-and-trade scheme, one individual's choice to increase their emissions would require them to purchase someone else's permits, resulting in no overall increase in atmospheric greenhouse gas levels. These examples don't exhaust the range of possibilities, but they illustrate how the practical upshot of diminishing environmentalists' moral motivation depends on how else climate change is being addressed.

Yet, note also that my claim it's okay to focus on problems other than climate change is only one side of the argument I advance in this book. The other side is that the climate movement can position itself for greater success by embracing its role in the social division of labor. Under the status quo, people generally do not constrain their contributions to climate change to levels that approximate what's needed to eliminate the problem. I see no reason to expect this to change in the foreseeable future through efforts to impose moral pressure. By reorienting themselves toward recruiting and retaining contributors who are not obligated to stay involved, members of the climate movement stand to offset losses of existing efforts with new contributions. By prioritizing strategies that modify incentives and structural features of individuals' choice environments, climate activists also may be able to alter the behavior of vast numbers of people who currently do not contribute to their cause. And by stepping back from rigid behavioral prescriptions and polarizing condemnations of popular lifestyles, the climate movement may be able to relax obstacles that currently impede its success.

For her part, Hourdequin is friendly to my suggestion that climate advocates should abandon their most narrowly prescriptive demands. However, she wonders whether at least some of the power of a moralized approach can be retained with more flexible norms. Specifically, Hourdequin proposes a norm that affirms "almost everyone has reasons and opportunities to do something about climate change" (p. 112, this volume). By holding people accountable for their responses to these reasons and opportunities,

Hourdequin suggests it may be possible to achieve better results than by giving up on norms altogether.

As I said in my earlier reply to Hourdequin's opening statement, I am skeptical that talk of "reasons and opportunities" can do the work Hourdequin wants it to do. Although I agree these reasons and opportunities exist, I deny this implies people have decisive grounds to prioritize climate action over other outlets for altruism. Since I've already developed this point at length, I won't belabor it here.

Yet, Hourdequin's talk of a moralized norm raises a further difficulty for her proposal. In most scholarship on norm-based cooperation, norms are understood as prescribing or deprecating specific behaviors (e.g., Ostrom 1990; Ellickson 1991; Bicchieri 2006). This is not an accident: the social pressure that gives norms their power depends on monitoring and enforcement by members of the norm-following community. Especially in large societies where individuals do not know each other well, it is infeasible to monitor and enforce norms that make no specific behavioral demands. This is why norms in large-scale communities typically revolve around patterns of behavior others readily can observe.

Thus, although Hourdequin's flexible formulation in terms of "reasons and opportunities" avoids many problems that come with narrow behavioral prescriptions, this flexibility makes her proposed norm a poor candidate for monitoring and enforcement at large scales. Except in cases where people have intimate familiarity with others' lives and thought processes, it would be impossible to assess whether a given person was responding appropriately to the norm. Hence, we could expect the norm to face one of two practical fates. On the one hand, it could simply fail to sustain compliance, since deviants might face little pressure to change their behavior in any specific area of life. On the other hand, the norm might encourage the "mere window dressing" Hourdequin laments, whereby people make token efforts to convey they are "doing something" without investing meaningfully in climate action.

With this said, it bears noting a norm along the lines Hourdequin describes could have greater chances of success in settings other than the broad social context. For example, in small, intimate groups, it may be feasible to uphold mutual expectations that reinforce diverse but substantial responses to climate change. Because of their limited reach, these norms likely would have modest impacts, and we should expect them to arise primarily among people who are independently committed to fighting climate change. But even today, it already is possible to observe moralized norms among environmental activists that sustain climate action effectively.

Hourdequin's discussion of the Paris Agreement highlights another possible application of a flexible norm-based approach. As I explained earlier in this book, the Paris Agreement does not codify specific responses to climate change; rather, it enjoins nations to "undertake and communicate ambitious efforts" to respond, leaving parties to determine these responses' details themselves (United Nations 2015). Hourdequin introduces the Paris Agreement as a cautionary tale since it has not yet yielded the decisive responses many would like to see. But it bears noting that the global diplomatic arena has features that make it more promising as a venue for Hourdequin's approach than society at large. There are relatively few nations (the United Nations recognizes 193 member states; United Nations 2023a), making it possible to achieve rich global discourse of a kind citizens cannot approximate among themselves. Nations also have resources to engage in very different forms of measurement, reporting, and oversight than private citizens can muster.[6] Because of factors like these, we can imagine how flexible, norm-based cooperation among nations could sustain diverse but ambitious climate efforts.[7]

The case for pursuing a flexible, norm-based approach in global diplomacy arises from features the diplomatic arena shares with small, intimate groups. However, by the same token, I believe reflecting on the differences between United Nations deliberations and decentralized social interactions should lead us away from Hourdequin's approach in the latter context. In large, anonymous communities, a norm that urges responsiveness to "reasons and opportunities to do something about climate change" (p. 112, this volume) – but does not demand or prohibit anything specifically – seems destined for ineffectiveness or failure.

5.   A Worthy Option

Because of this book's presentation as a debate, much of my contribution has been negative, arguing it can be okay to focus one's altruism on issues other than climate change. However, I also have tried to emphasize my position's positive upshot that those who pursue climate action should proceed differently than would make sense if everyone had a duty to join. Happily, Hourdequin's comments echo many of my views on this score. Thus, in closing, I want to highlight this area of constructive convergence.

For all the climate movement's talk of climate change as "the greatest threat the world has ever faced" (UNHCR 2022), the movement historically has struggled to identify and communicate compelling ways for people to act on the problem. By way of illustration, consider the United Nations' list of ten "actions for a healthy planet" published as part of its recent Act Now campaign:

1. Save energy at home.
2. Change your home's source of energy.
3. Walk, bike, or take public transport.
4. Switch to an electric vehicle.
5. Consider your travel [i.e., take fewer airplane flights].
6. Reduce, reuse, repair and recycle.
7. Eat more vegetables.
8. Throw away less food.
9. Plant native species.
10. Clean up your environment.
11. Make your money count [i.e., purchase green products and invest in sustainable businesses].
12. Speak up [i.e., raise awareness, encourage others to act, and appeal to political leaders].

(United Nations 2023b)

Take a moment to imagine someone who does all these things – call him Harvey. Harvey's home is muggy in summer and chilly in winter. He air-dries his clothes. He cleans his dishes in a washbasin with biodegradable soap and dumps the gray water into his native plant garden. He generates most of his electricity from rooftop solar panels and times his laundry to run when they are producing. He walks or bikes whenever possible and resists traveling by plane. He does not own a car. He avoids new purchases and mends worn-out clothing by hand. He recycles everything his city will accept, including things he needs to drop off in person. He eats a vegan diet and carefully portions his meals to prevent waste. He maintains a healthy compost heap. He participates quarterly in community trash cleanups. He shops selectively at green businesses and invests his savings in an environmentally oriented mutual fund. He rarely gets through a meal or social gathering without bringing up the climate crisis. He supports green candidates passionately. He is on a first-name basis with local officials after a long history of outspoken advocacy.

To say the least, Harvey is trying hard to make the world a better place. Yet, two things should strike us about him as an exemplar of climate action. First, Harvey's actions likely will seem burdensome and unappealing to many people who might be open to acting on climate change. This is not just because Harvey gives so much of himself to his cause. It is also because his lifestyle differs dramatically from what others consider "normal" and because virtually nothing he does produces a perceptible positive impact on the world or any specifiable person. (The main exceptions are Harvey's trash cleanups and native plant garden, which are noteworthy for having arguably the least to do with climate change of anything on the list.) Second, Harvey's efforts make no evident use of his distinctive talents, experiences, and social relationships. They are utterly generic – the sorts of things *anyone* could do – and thus plausibly waste the special attributes *Harvey* brings to the table.

Despite these drawbacks, it could make sense for the climate movement to elevate Harvey as a role model if certain things were true about the theory and practice of fighting climate change. For instance, if morality required people to restrict their ecological footprints and promote political action on climate change, activists could shrug their shoulders and insist we *must* become like Harvey if we wish to be good people. Similarly, if the only sensible way to address the climate crisis were to pressure people into behaving like Harvey, the case for highlighting his conduct would be clear. Yet, I have disputed both possibilities. There is no specific duty to become like Harvey, and there is no specific need to tackle climate change by enforcing behaviors like his. Harvey's efforts represent only one approach among many for discharging his duty to engage in pro-social action. Given this, Harvey seems less like an exemplar to elevate than like a branding liability.

How can the climate movement respond to the issues that make Harvey a problematic spokesperson? First, instead of focusing on actions likely to strike many people as unappealing, weird, and ineffectual, the movement could highlight opportunities that are engaging and rewarding. To recall an earlier example, planting trees around one's neighborhood or workplace can advance the climate change cause in many ways including by sequestering CO_2, reducing urban heat island effects, shading pedestrians and nearby buildings, and enhancing wildlife habitat. Moreover, this activity is a fun and uplifting way to bring people together outdoors, beautify one's community, and create a tangible memory that will last and grow over decades.

Consider the impact someone could have by investing in urban tree planting with the same dedication and drive we ascribed to Harvey. Is it obvious such a person would be *less* useful to the climate movement? Is it obvious such a person would be *less* successful at inspiring others to join her? I would expect the opposites to be true. Although this is only one example, the point generalizes easily. By attending closely to what types of activities inspire and reward enthusiasm, the climate movement stands to increase its impact even while letting go of the fantasy of broad compliance with a specific list of behaviors.

Climate activists also would do well to think more about how diverse forms of participation may suit different individuals. Instead of offering generic lists of actions anyone can take, climate communicators can challenge people to reflect on their backgrounds and interests and then help them connect with activities that fit *them*. Possibilities for tailored engagement are easy to imagine, and Hourdequin's discussion is rich with possibilities along these lines. To mention some additional prospects: Coastal residents can participate in local preparations for sea level rise. People in wooded areas can help thin forests to reduce vulnerability to fire. Animal lovers can enhance migration corridors to help wildlife respond to changing conditions. Promoters of social justice can work with disadvantaged populations to ensure they are not excluded from adaptation planning. Citizens with strong rhetorical skills or political connections can seek headway in the policy arena. Exploring these options fits well with Hourdequin's discussion of polycentric climate solutions, where individuals and groups at different levels play diverse roles in tackling the problem. Not only would these varied forms of involvement use people's capabilities and passions more efficiently, but the improved fit between individuals and actions also would foster deeper and longer-term engagement.

As I hope these comments illuminate, the upshot of the position I've defended is not simply that we may decline to act on climate change. It's also that reorienting toward recruitment and retention can position the climate movement for greater success. It's an empirical fact as well as a theoretical one that people have diverse options for allocating their altruistic energies. The climate movement must compete with many other outlets for individuals' limited band-width. By offering more attractive engagement opportunities and economizing on contributors' efforts, the climate movement may be able to increase its efficacy even while backing away from its universalist demands.

To the extent such efforts succeed, the practical distance between Hourdequin's position and mine will narrow. Even if there's no *duty* to devote one's altruism specifically to fighting climate change, it would be difficult to resist joining a climate movement that offered varied options for inspiring, rewarding, high-impact participation. For those of us who care deeply about this issue, let us therefore build toward a future where more people help tackle climate change, not because they feel obligated to do so, but rather because it's their best opportunity to help make the world a better place.

Notes

Foreword

1   The +2°C goal was first articulated in the 2009 Copenhagen Accord and reaffirmed in the 2010 Cancun Agreement. The 2016 Paris Agreement objective is to hold long-term global average temperature increase "well below 2°C above preindustrial levels and pursuing efforts to limit the temperature increase to 1.5°C above pre-industrial levels."

2   Hourdequin spends considerable time pushing back against the idea individual emissions don't matter morally. Although the emissions of an individual action may be inconsequential, she follows John Nolt to suggest one's cumulative lifetime emissions may not be causally inconsequential. Furthermore, individual emission choices are entangled with "the complex politics of environmentalism and climate change" so they have political meaning "that reverberates beyond their effects on the climate system." Reducing personal emissions, then, can be a part of building a culture of environmental sustainability.

Opening Statement – Hourdequin

1   See   www.epa.gov/system/files/documents/2023-04/US-GHG-Inventory-2023-Main-Text.pdf, figure ES-1.

2   There is some empirical evidence supporting this claim. For example, Nassauer, Wang, and Dayrell (2009) found that Michigan homeowners' landscaping preferences were strongly influenced by their neighbors' yard designs. See also Goddard, Dougill, and Benton (2013), and a review by Cook, Hall, and Larsen (2012), which suggests that residential landscaping choices are influenced by multiple factors at multiple scales.

3   A tragedy of the commons or intergenerational moral storm analysis might be apt in describing certain ways in which people and nations *have responded* to climate change, but it is not, I believe, apt in describing how they should. On this point, Stephen Gardiner and I agree: he finds that people often misinterpret his analysis as suggesting that we are

stuck with the moral storm and with self-interest as the main driver as climate policy. That is not his view; instead, he argues that institutional transformation is needed to incorporate ethical values that are not embedded in current social, political, and economic systems.

4   See www.sunypress.edu/p-2427-watsuji-tetsuros-rinrigaku.aspx

5   This characterization of ethics strikes me as important because I think it resonates with a conception of ethics that is implicit in the understanding of many, yet captures an aspect of ethics that is undertheorized, particularly in the highly individualistic Western societies Watsuji aimed to critique. So an ethical focus on betweenness, in my view, is relevant not only to a particular cultural context, but has the potential to gain traction and bear fruit more broadly. One way of drawing out the relevance of this relational approach is by connecting it to virtue ethics, as I do here. As my particular focus is on environmental ethics, Watsuji's explicit inclusion of both social and natural environments in his understanding of betweenness provides a helpful starting point for the investigation.

Opening Statement – Shahar

1   The $2.70 figure adjusts the $1.90 benchmark (given in 2011 dollars) for inflation to 2025 using the U.S. Inflation Calculator available online at www.usinflationcalculator.com.

2   Note that if the European Union is considered as a single entity rather than as a union of separate countries, it qualifies as the third largest emitter, moving India into fourth place.

3   As you've seen in Hourdequin's opening statement, I am not the first to deny people are obligated to tackle climate change unilaterally (for other arguments in this vein, see Johnson 2003; Sinnott-Armstrong 2005; Sandberg 2011; Kingston and Sinnott-Armstrong 2018). As will become apparent, my position differs from those of these earlier authors in virtue of its focus on a division of labor in responding to serious problems.

4   The line of argument presented in the next two sections is adapted from an earlier paper (Shahar 2016).

5   The absence of meaningful market mechanisms creates an important challenge for coordination among altruists since individuals cannot rely on price signals to direct their efforts where they're needed most. On prices' role in facilitating economic coordination, see Hayek (1945). Figuring out how to channel one's altruism effectively is an important puzzle I touch on below. For another noteworthy approach to tackling this puzzle, see Gertler (2021).

6   For further discussion of this example, see Schmidtz and Willott (2006).

7   For discussion of a similar scenario from which this example was adapted, see Sinnott-Armstrong (2005, pp. 289–290) and Raterman (2012, pp. 424–425, 427–428).

8 Along related lines, Ryan Flugum and Matthew E. Souther (in press) offer empirical evidence that companies strategically emphasize environmental, social, and governance (ESG) commitments to conceal poor performance in more conventional financial categories.

9 For discussion of this problematic tendency in other areas of environmental discourse, see Shahar (2024).

10 Many people believe it's wrong to eat meat independent of the climate change problem. For discussion of this issue, see Shahar (2022).

11 All the drawbacks discussed below concern the *consequences* of trying to cultivate norms around climate-impacting behaviors. Different ethical theories disagree about how consequences fit into moral evaluations, and I don't want to imply morality always boils down to achieving good results. Still, given climate change's seriousness, I expect every plausible moral theory would concede the importance of adopting measures with a high likelihood of beneficial impacts. Even if good consequences aren't the *only* things that matter, they're certainly *among* the things that matter, and their importance is especially central when tackling something as harmful as climate change.

12 For a brief review of varying interpretations of the idea of social norms, see Constantino et al. (2022, pp. 57–59).

13 For discussion of why a one-size-fits-all approach may sometimes be desirable despite its reduced efficiency, see Mayer (2019).

14 In the case of positive incentives rewarding beneficial behaviors, the increase in cost comes in the form of higher opportunity costs for alternatives that are not rewarded.

15 Less salient but also noteworthy is the fact a norm-based approach would give Prius drivers little incentive to further restrict their car emissions. By enabling these individuals to be seen as good norm-followers regardless of their travel decisions, climate norms might miss worthwhile opportunities to push these drivers toward even less impactful patterns of behavior.

16 This same reasoning helps explain why norms often work so well in other environmental domains such as preventing littering or encouraging recycling. Although norms against littering or trashing recyclables may not be sensitive to differences in individuals' circumstances and values, there is typically little risk of being overly strict and imposing needlessly high costs on the way to producing desired environmental outcomes.

17 See in this vein Kahan (2010) and Gromet, Kunreuther, and Larrick (2013). For a striking illustration of alienation from stereotypically "green" behaviors, see Tabuchi (2016). Importantly, the conformity pressure generated by anti-environmentalist alienation may extend beyond those who feel personally alienated, also affecting individuals who seek approval from such people (see, e.g., Brick, Sherman, and Kim 2017).

18 For one illustration of taking climate change seriously while honoring sensibilities that conflict with mainstream climate norms, see Cloud (2016).

19 It's especially clear a candidate's stance on climate change shouldn't automatically determine our vote in elections for offices with little authority over climate policy.

20 At the 2023 Conference of the Parties to the Paris Agreement at which the "global stocktake" took place, delegates made bold-sounding commitments to contribute to "Transitioning away from fossil fuels in energy systems, in a just, orderly and equitable manner, accelerating action in this critical decade, so as to achieve net zero by 2050 in keeping with the science" (UNFCCC 2023, p. 5). However, in keeping with the Paris Agreement's general structure, each nation was left to determine the details of these contributions independently, "taking into account the Paris Agreement and their different national circumstances, pathways, and approaches" (UNFCCC 2023, p. 5). Although the conference participants hailed their efforts as a historic success, other commentators have been less effusive, with one noted climate scientist opining, "It was weak sauce. What we really need is a commitment to phase out fossil fuels, on a very specific timeline: We're going to reduce carbon emissions by 50 percent this decade, bring them down to zero mid-century. Instead, they agreed to transition away from fossil fuels – the analogy that I use is, you're diagnosed with diabetes, and you tell your doctor you're going to transition away from doughnuts. That's not going to cut it. It didn't meet the moment" (Michael Mann, quoted in Gramling 2023; see also Pearce 2024).

21 Even if we're motivated to promote good causes by altruism rather than self-interest, that doesn't make self-interest irrelevant to which causes and actions we should choose. If two causes are equally good, it's sensible to pick the one that would benefit us more. If one cause is only a little less impactful than another but significantly more beneficial or us, that could also be a valid reason to pick it, especially if this allowed us to justify devoting more time and resources to it.

Reply to Hourdequin

1 I say "simplistic" because there are more sophisticated ways to think about consequentialism that support very different forms of moral reasoning. For instance, a thoughtful consequentialist might ask whether society would be better off if people developed strong commitments to integrity like Hourdequin favors and, if so, claim people should develop such commitments. See along these lines Jamieson (2007).

2 Broome's estimate was published in 2012, so these numbers would be higher now due to inflation.

3 As the Intergovernmental Panel on Climate Change explains, "A dynamical system such as the climate system, governed by non-linear deterministic equations, may exhibit erratic or chaotic behavior in the sense that very small changes in the initial state of the system lead

to large and apparently unpredictable changes in its temporal evolution. Such chaotic behavior limits the predictability of the state of a non-linear dynamical system at specific future times, although changes in its statistics may still be predictable given changes in the system parameters or boundary conditions" (IPCC 2021, p. 2221).

4   It's a point of mathematics that if you divide a positive quantity of harm by a positive number of individuals who contribute to the harm, this will yield a positive average quantity of harm per contributor.

5   For criticism of Nolt's methodology, see Hartzell (2011), Odenbaugh (2011), and Sandler (2011).

6   Nolt reports the figure as 1,084 metric tons, which is approximately equivalent to 1,194.91 U.S. tons.

7   The EPA's figure is 8,887 grams of CO_2 per gallon of gasoline, which is approximately equivalent to 19.59 pounds. In practice, it's plausible that a person's decision to purchase one gallon less of gasoline would make no difference to the total amount of gasoline produced and sold in the market. I leave this detail to the side for the sake of simplicity.

8   For further discussion of why integrity-based reasoning is a valuable resource for driving climate action in the absence of individual efficacy, see Jamieson (1992).

9   In some places in her discussion, Hourdequin qualifies this claim to apply principally to people who recognize climate change's reality and seriousness. I will not belabor this qualification since my discussion takes it for granted that climate change is real and serious.

10   Note that some writers advance similar rationales to the one Hourdequin presents to argue people of integrity will avoid consuming factory farms' products. I examine these perspectives in Shahar (2022, ch. 6).

11   The canonical articulation of this characterization of environmental problems is in Hardin (1968). For critical discussions of how this model can be applied to climate change, see Gardiner (2006) and Brennan (2009).

12   It bears noting that authors who depict humans as purely self-serving agents often do so for theoretical purposes other than describing what actual people are like. For instance, by imagining agents who are motivated solely by monetary profits, scientists can introduce simplicity and determinacy into their models that would be difficult to replicate with more sophisticated psychological representations. Characterizing individuals as having specific motivations can also be useful for exploring limited domains of activity in which those motivations dominate, even if the resulting theories are inapplicable outside those domains. For classic examples of these types of theoretical decisions, see Mill (1844) and Wicksteed (1910, pp. 158–211). For discussion, see Gaus (2008, ch. 1).

13   Stated more carefully: I've been disputing whether people *typically* have decisive reasons to participate in climate action. More on this qualification below.

Closing Statement – Hourdequin

1   It is tempting to say "everyone" here, or "all moral agents," but it is important to acknowledge that not *everyone* is well positioned to make the world a better place in the way Shahar envisions. People who have very limited resources and need to focus primarily on meeting day-to-day needs may have limited capacity to take up a further cause, for example, though notably many people in situations of extreme duress *have* taken steps to make the world a better place. Regardless, what one can and should do may depend importantly on one's positionality and one's values, as well as time, money, power, and one's social and institutional relationships.

2   My friend's intervention was indirect in the sense that she did not personally alter the intersection; she engaged with decisionmakers who had the power to direct the changes.

3   Ty Raterman (2012) uses a similar example. He says that one individual riding a jet ski has little effect on the overall pollution of a lake (Raterman 2012, p. 420), even though the collective effects of many people riding jet skis can be significant.

4   Interestingly, using an analogous example, Sinnott-Armstrong (2005) suggests that adding an additional quart of water to an already-flooding stream is not morally problematic: after all, the stream is *already flooding*. But adding water to an already-flooding stream or an overflowing bathtub makes the problem worse (even if adding a small increment of water makes the problem just a little bit worse). If the bathtub is overflowing, the first thing to do to limit the damage is generally to turn off the water.

5   There is a large literature on "we-reasoning," with diverse views of what it involves and its relation to ethics. For examples and further discussion, see Hakli, Miller, and Tuomela (2010), Tam (2020), and Schwenkenbecher (2019, 2021); for related discussion of "team reasoning," see Radzvilas and Karpus (2021). Literature on "we-reasoning" also intersects with work on shared/joint intentions, collective action, collective responsibility, and a nice overview of literature in these areas is provided by Smiley (2023).

6   Shahar might agree that we have relational responsibilities to one another, but he construes those responsibilities differently than I do. He would likely argue that because people's actions fit together to accomplish collective goals, we can divide up the work on those goals (Shahar, this volume and pers. comm.). People need healthcare, but not everyone needs to be a doctor. People need food, but not everyone needs to be a farmer. Similarly, we need to address climate change, but not *everybody* needs to work on climate change. I have argued, however, that because climate change is a fundamental and cross-cutting problem, we have reasons to consider how actions in many domains, from individual to institutional, can contribute positively to addressing it. However, my view is consonant with Shahar's in the sense that I believe that we can divide up the labor of responding to climate change, taking advantage of the particular skills, expertise, capacity

and influence each of us has in our respective communities. Which view is correct depends partly on empirical questions about the extent to which a group of specialists can adequately address climate change while others focus their attention elsewhere, but we also view the ethical issues differently, as I think that most people have reasons to act with awareness of climate change, just as they have reasons to act with awareness of social issues like racism and sexism. That doesn't mean that some of us can't focus intensively on climate change while others focus less intensively, and in this sense, I accept that some degree of specialization is appropriate and beneficial.

Closing Statement – Shahar

1   It may be worth adding that although I believe each of us must tackle some of the world's problems, I also believe valid contributions can take many forms including some that look different from "activism" or "charity" as commonly conceived. For example, it sometimes has been observed that leaders in private enterprise stand to benefit society at least as much by devoting themselves to their work as through overt altruism (see, e.g., Smith [1776] 1904, IV.II; Friedman 1970; Brennan 2012). To the extent this is so, it might be possible for at least some highly productive individuals to fulfill their pro-social obligations by increasing their efforts at work rather than through recognizable forms of altruism. Of course, since such efforts often yield benefits like financial remuneration, professional advancement, and career satisfaction, we may question whether specific individuals truly are sacrificing when they eschew overt altruism to invest more in their careers. But if someone sincerely judges the most effective way to help tackle the world's problems is to concentrate their sacrifices in the office, I see no theoretical grounds to condemn this choice. Although people who serve society this way may *look like* they are "specializing in self-interest," what sets their efforts apart are the sacrifices they make to help tackle the world's problems.

2   Arguably, we still would have fairness-based duties in such a scenario to ensure the superheroes did not have to carry the burden by themselves. However, if the superheroes enjoyed their roles and could solve our problems without sacrificing anything, it seems doubtful we would have an obligation to tackle problems at a substantial cost to ourselves.

3   For a canonical expression of this point, see Kant ([1797] 1996, IV.VII–VIII).

4   For a classic discussion of this point, see Sen (1981).

5   Immanuel Kant, the philosopher arguably most concerned with autonomy in the mainstream canon, certainly conceived of autonomy in this way. See Kant ([1785] 2006).

6   As I explained in my opening statement, this is not to say nations have succeeded at establishing effective systems of monitoring and enforcement. The point is simply that it is conceivable for nations to measure,

report, and oversee their respective contributions to global climate efforts in a way that would be unimaginable for citizens acting on their own. It also is worth emphasizing that there are forms of monitoring and enforcement that are feasible for citizens and not nations. However, these operate primarily at small scales and are not as useful for achieving global coordination.

7   In practice, it seems arguable the Paris Agreement's disappointing achievements result from a combination of the noncompliance and "mere window dressing" described above. If this diagnosis is apt, it would illustrate that effective norm-based cooperation is far from guaranteed even among nations.

Further Reading

For general introductions to climate ethics, readers may wish to consult Stephen Gardiner's *A Perfect Moral Storm: The Ethical Tragedy of Climate Change*, and the more recent and very accessible *Dialogues on Climate Justice* by Gardiner and Arthur Obst, which includes a chapter on individual responsibility. Elizabeth Cripps's *Climate Change and the Moral Agent: Individual Duties in an Interdependent World* is an important scholarly discussion of individual duties in relation to climate change, and it offers a perspective somewhat different from those developed here.

Further Readings Related to Hourdequin's Contribution

The position I develop in this book emphasizes integrity, relationality, sociality, and interactions at multiple scales as important to understanding individuals' roles and responsibilities in responding to climate change. On integrity, Audi and Murphy's "The Many Faces of Integrity" is a helpful starting point. With respect to relationality, there are resonances between my approach and Iris Marion Young's social connection model of responsibility, which offers a way of conceptualizing individual responsibility for social injustice. Young's model is developed in the book, *Responsibility for Justice*, and in the shorter piece, "Responsibility and Global Justice: A Social Connection Model." Her view suggests that individuals who contribute to large-scale injustices – such the exploitation of sweatshop workers who produce clothing to address global distribution – bear forward-looking responsibility for structural and systemic injustices. She further emphasizes the importance of individual responsibilities to work with others to effect systemic and political change. Responsibilities in turn depend on factors such as one's social position, privilege, and access to power. Eric Godoy applies Young's social connection model to climate change in his helpful article, "Sharing Responsibility to Divest from Fossil Fuels."

My approach is arguably less explicitly political than Young's, suggesting that individual actions to address climate change – even outside of explicit efforts to organize politically – have the potential to shape the thinking and actions of others, the development and transmission of norms, and the cultivation of cultures that support mutual flourishing over time. This approach is rooted in a relational conception of persons and of ethics. A nice introduction to relational ethics can be found in Thaddeus Metz and Sarah Clark Miller's chapter, "Relational Ethics," and an article I co-authored with David B. Wong, "A Relational Approach to Environmental Ethics," develops one approach to grounding environmental responsibilities relationally. My "Climate, Collective Action, and Individual Ethical Obligations" provides an earlier and more condensed version of the view developed here, focusing on integrity and relationality as important to understanding individual climate responsibilities. Finally, for an important perspective on the relational dimensions of climate justice, see Kyle Whyte's "Too Late for Indigenous Climate Justice: Ecological and Relational Tipping Points," which emphasizes the importance of relational qualities such as trust, accountability, and reciprocity in responding ethically to climate change.

Further Readings Related to Shahar's Contribution

The idea at the heart of my contribution to this book is that specialization and division of labor make sense in the context of altruism, just as they do in the economic arena. For a canonical treatment of this idea in its original context, it's hard to do better than Adam Smith in chapter 1 of *An Inquiry into the Nature and Causes of the Wealth of Nations*. Smith's account remains essential reading for any student of economics, and I would argue the same is true for altruists as well.

Beyond this central idea, many arguments throughout my portions of this book build upon a baseline understanding of how economic and political theory bear on ethical matters. For an excellent introduction to the intersections of these disciplines, see Gerald Gaus and John Thrasher's *Philosophy, Politics, and Economics: An Introduction*. On social norms and informal cooperation in particular, see Cristina Bicchieri's *The Grammar of Society* and Elinor Ostrom's *Governing the Commons*. All three are a bit technical, but those who seek more rigorous understanding of social interactions will find the investments worthwhile.

If the position I defend in this book is correct, each of us faces difficult choices about how best to allocate our altruistic energies. For influential contemporary attempts to help people navigate these choices, see William MacAskill's *Doing Good Better*; Peter Singer's *The Life You Can Save*; and Benjamin Todd's *80,000 Hours: Find a Fulfilling Career That Does Good*.

Glossary

Altruism: This term refers to a commitment to make the world a better place, even at a cost to oneself, as well as actions taken based on such a commitment. Correspondingly, an "altruist" is a person who is committed to altruism and acts accordingly.

Attribution Science: An area of scientific research focused on understanding the role of climate change in causing or exacerbating extreme events such as hurricanes or floods.

Backward-looking Responsibility: A conception of moral responsibility focused on accountability for past wrongs. Backward-looking responsibility is often contrasted with forward-looking responsibility, which emphasizes the capacity for contributing positively to justice in the present and future.

Carbon Footprint Analysis: An approach to measuring emissions at the individual or institutional level – an agent's "carbon footprint" – typically by estimating and adding up emissions in various domains. Individual carbon footprint analyses often involve inputs such as distance commuted to work or school, type of diet (vegan, vegetarian, omnivorous), the size of an individual's dwelling, etc.

Chaotic Systems: A chaotic system is one in which it is difficult to make detailed predictions about future states of the system even though the system's elements behave in an orderly way. A system's chaotic character can arise from diverse sources. However, one source that is relevant to this book's discussion comes from elements in a system displaying sensitivity to small changes in how other elements behave. When systems contain many interdependent elements that are sensitive to these small changes, minor alterations to the system can have cascading effects that are difficult to anticipate with much granularity. A popular representation of this type of phenomenon is termed the "butterfly effect," referring to the idea that the flap of a butterfly's wings in one part of the world might initiate a cascade of small changes throughout the atmosphere that ultimately results in a major event on the other side of the globe, such as initiating a tornado that would not otherwise have occurred. (For the original source of this metaphor, see Lorenz 1972.)

Complicity: This term refers to a condition of being "involved" in a wrongful act – even if one did not directly perform the act oneself, cause it to occur, or meaningfully increase the likelihood of its occurrence. In the context of this discussion, complicity is significant precisely because it is possible to be complicit in wrongdoing even while having no direct impact on outcomes that occur.

Consequentialism: This is an umbrella term for a family of ethical theories that seek to account for the moral significance of actions (e.g., their goodness and badness, rightness and wrongness) by appealing to the moral value of outcomes with which they are associated. Correspondingly, "consequentialist" arguments, inferences, or considerations are those that appeal to consequences or outcomes to ground ethical claims. Although all forms of consequentialism focus on outcomes, it is important to stress that consequentialist theories can vary dramatically in which outcomes are highlighted as important and how these should be considered. For instance, it counts as a form of consequentialism to say each individual act should be evaluated based on its impacts on humans' psychological well-being, and it also counts as a form of consequentialism to say societies' rules and thought patterns should be evaluated based on their long-term impacts on people and the natural world. These examples illustrate that consequentialism does not necessarily imply any specific set of moral judgments; rather, it refers to a more fundamental orientation toward how we should approach moral questions.

Coordination: In the context of social behavior, coordination refers to the alignment of different individuals' actions. Coordination is especially significant in contexts where individuals cannot bring about certain outcomes on their own and can do so only with contributions from other individuals. Fostering coordination is a vital function of public policies, social norms, private agreements, and numerous other mechanisms of inter-personal control.

Costly Unilateral Action: In the context of social interactions, a unilateral action is one taken by an individual without others' agreement and without assurance that others will respond cooperatively. A *costly* unilateral action is one that involves a burden or sacrifice on the part of the actor. In the context of this book, costly unilateral actions are significant because they are often taken by altruistic individuals who wish to promote desirable social outcomes; yet their impacts often depend on whether others follow suit, and this is not always likely. (By way of contrast, see "Coordination" above.)

Division of Labor: In economics, division of labor occurs when complex tasks are decomposed into simpler sub-tasks and allocated among multiple individuals, each of whom focuses on only a portion of the overall work to be performed by the group. As Adam Smith famously celebrated in *An Inquiry into the Nature and Causes of the Wealth of Nations* (Smith [1776] 1904, I.I.), division of labor can facilitate dramatic increases in productivity for a variety of reasons. Some of these reasons are mundane, such as eliminating the time wasted when switching between tasks. But some are vitally important, such as allow-ing individuals to become specialists in particular areas and increase chances for individuals to identify opportunities to innovate in their chosen fields. As Smith observed, when indi-viduals specialize in diverse tasks and combine their efforts, they can often achieve output many times as great – and of much higher quality – than the same number of people who each tried to accomplish the same objectives independently. In this book, the ideas of specialization and division of labor are extended into the domain of ethics and altruism. The division of *moral* labor occurs when people pursue ethical objectives like alleviating poverty, advancing justice, and protecting the environment through a strategy of specialization and division of labor rather than by having each person pursue a wide range of objectives. On this approach, the complex task of tackling the world's problems would be decomposed into simpler sub-tasks and allocated among many individuals, each of whom would focus on only a portion of the overall work to be performed by the group. If the effects of specialization and division of labor in this context were similar to what they are in other contexts, this would imply that people who pursued a division of moral labor could accomplish much more together than if they each tried to tackle many problems independently.

Economizing: In *An Essay on the Nature and Significance of Economic Science,* Lionel Robbins memorably defined economics as "the science which studies human behavior as a relationship between ends and scarce means which have alternative uses" (Robbins 1932, p. 15). Although this is not the only way people have sought to characterize the subject matter of economics, Robbins' discussion helped popularize the idea that economics is concerned fundamentally with finding the best ways to allocate scarce resources among competing ends. In line with this idea, the notion of "economizing" refers to the judicious use and management of resources to ensure they go to their highest value uses. Since the purpose of economizing is to maximize benefits from a limited budget, economizing is not the same as using assets as little as possible. Rather, the idea is to allocate resources rationally across their alternative uses to ensure they are not squandered on low value activities while leaving more desirable goals unmet.

Hypocrisy: Hypocrisy is a way of acting or speaking that shows inconsistency between word and deed, or between holding oneself and others accountable. Examples of hypocrisy include professing a great commitment to respect while consistently treating others disrespectfully, or criticizing others for dishonesty while lying and cheating regularly oneself.

Incentive: An incentive is anything that motivates someone to do something. For example, if I offer you $20 to do a job for me, the prospect of receiving the $20 would give you an "incentive" to do the job. (In some cases, a specific incentive might fail to generate a desired action – for instance, because the incentive was not strong enough to overcome other factors weighing on the person's decision. For example, if working for me would cause you serious harm, the $20 incentive likely would not be enough to induce you to accept my offer.) In this book, we will often speak of "incentives" not just in terms of specific inducements but also in terms of the full spectrum of motivators bearing on a person's decisions. In this sense, offering you

$20 would modify your "incentives" by attaching additional benefits to the option of working for me.

Inefficacy: To call something "efficacious" is to say it has the power to produce a specific result. On the flip side, something that lacks this power is "inefficacious" with respect to the result in question. In the context of this book, we will talk about "inefficacy" primarily in connection with a person's ability to meaningfully influence the severity of global climate change (see also "Problem of inconsequentialism").

Integrity: A virtue related to the ideas of coherence or wholeness. Integrity involves the relationship between values and actions, and relationships among diverse values. A person of integrity seeks coherence between their values and actions and works to harmonize diverse values (recognizing that values sometimes come into conflict). This enables a person's values to fit together into a coherent whole, and to find expression throughout their life.

Lottery Ball Machine: This is a device used to facilitate random selections. A typical lottery ball machine features a chamber filled with hollow plastic balls, each marked with a specific number. As air is blown through the chamber, the balls bounce around wildly. Balls are selected by opening a tube on the side of the chamber and allowing the fan to blow them out.

Luxury Emissions: Greenhouse gas emissions in excess of what is needed to live and flourish, such as emissions associated with plane travel for vacations or second homes.

Moral Corruption: A tendency to distort moral thought and discourse in self-serving ways, or as Steve Gardiner puts it, "the subversion of our moral discourse to our own ends" (Gardiner 2010, p. 286). In the case of climate change, moral corruption may lead current generations to postpone significant action and to make decisions that do little to address the problem, while justifying these choices through mechanisms such as self-deception.

Moralization: This is when people impute moral significance to things that otherwise could have been seen as morally neutral. To illustrate, consider the difference between a "fee" and a "fine" for engaging in a specific behavior. In both cases, a person who acts in the relevant way will be required to pay a sum of money. But framing this in terms of a "fee" implies no moral judgment toward the behavior in question, whereas calling it a "fine" implies the behavior is being condemned. Patterns of moralization often reflect substantive judgments about what types of behavior are acceptable vs. unacceptable. However, in the context of this book, we also consider whether certain forms of moralization are socially useful or counterproductive – and whether it is wise for climate activists to use moralization as a tool for achieving their aims.

"No Regrets" Climate Action: In its Second Assessment Report on the economic and social dimensions of climate change, the Intergovernmental Panel on Climate Change (IPCC) memorably highlighted "no regrets" options for tackling climate change which would be worthwhile even if their climatic benefits were ignored (IPCC 1996, p. 15). The IPCC's point in highlighting these measures was to emphasize that regardless of one's perspective on the wisdom of bearing costs to combat climate change, it would be irrational to reject these "no regrets" opportunities. Although the IPCC's original invocation of this concept was pitched mainly at the level of public policymaking, this book extends the concept of "no regrets" climate action to the individual level to encompass options people have to act on climate change without incurring any costs to themselves.

Norms: See "Social norms."

Polycentric: Literally, this term means "multi-centered," and refers to approaches that seek to address collective action problems or other social issues at multiple scales. Polycentric approaches to climate change focus on actions and policies at multiple levels, seeking synergies between efforts at individual, local, regional, state, and international scales. According to Elinor Ostrom (2009a), polycentric approaches enable actors at different levels to experiment with various problem-solving strategies and to learn from one another.

Problem of Inconsequentialism: This is the challenge of justifying individual emissions reductions in response to climate change, given the apparently "inconsequential" impacts of individual emissions (or emissions reductions) in relation to the scale of the problem. The challenge emerges because climate change is the cumulative result of many relatively small contributions to the problem (see "Inefficacy"). In this context, individual actions to reduce emissions can seem inconsequential (and therefore difficult to justify), especially if others can't be relied upon to reduce their emissions as well.

Reasons of Integrity: Considerations that weigh in favor of certain actions in virtue of their connection to integrity. For example, a person for whom honesty is a key value has reasons of integrity (and perhaps other reasons) to act in accordance with that value.

Reasons of Relationality: Reasons for action that are rooted in our social connections with one another and to the broader world. Relational reasons emerge from connections with and responsibilities to others. They are tied to a broader conception of persons as relational, and of ethics as involving the character and quality of human relations with one another and with the broader world. Relational reasons can be grounded in particular relationships (e.g., people have relational reasons to care for their pets) or in more general relationships (e.g., as members of a broader community, people may have relational reasons to act in ways that contribute to the public good).

Social Norms: These are informal (i.e., unwritten, not legally enforced) rules that govern behavior in social settings. In *The Grammar of Society*, Cristina Bicchieri writes, "Norms refer to behavior, to actions over which people have control, and are supported by shared expectations about what should/should not be done in different types of social situations" (Bicchieri 2006, p. 10). For a brief review of varying interpretations of the idea of social norms, see Constantino et al. (2022, pp. 57–59). In all societies, social norms represent a key mechanism of interpersonal control that does not operate through official channels of public decision-making such as legislation, regulation, or jurisprudence. Despite their lack of formal political pedigree – and despite the sharp differences among social norms in different communities – compliance with social norms is often considered obligatory, and many public goods depend on such conformity. A key question throughout this book is whether the promulgation of new social norms should be seen as a core pathway for promoting action on climate change.

Structural Solutions: Many social problems can be traced not only to how individuals are disposed to act but also to features of the social environment in which those individuals carry out their affairs. When such "structural" conditions are important, it is often possible to make progress on alleviating problems by modifying the relevant features of the social environment. Because they transform how individuals' dispositions translate into specific actions, such "structural solutions" can often result in different patterns of behavior even if the relevant individuals remain disposed to act in the same way as before. To give a simple illustration, some traffic intersections are especially prone to collisions because they are characterized by confusing layouts, poor sight lines, or obstructed signage. In these cases, it may be possible to improve outcomes by addressing the structural features of these intersections that lead drivers to collide. Here, the key point would be that one can tackle the traffic safety problem without putting the onus on individual drivers to change their behavior. Even if drivers remain disposed to drive in the same way as before, a well-designed intersection will be safer because it channels these dispositions toward better outcomes.

Subsistence Emissions: Emissions needed to sustain basic life functioning, such as emissions associated with food, shelter, and other vital needs.

Tragedy of the Commons: This term was introduced by Garrett Hardin in 1968 to capture the tendency for valuable resource systems to become overused when many individuals desire to utilize them, there are no rules in place to control this utilization, and the total impacts of users' activities exceed the resource system's capacity to absorb them without harm (Hardin 1968). In such cases, Hardin argued it typically will be in each user's individual self-interest to expand their use of the resource system, even though the collective result of such behavior will be the destruction of the resource system. According to Hardin, the only way to avoid the inevitably tragic consequences of such self-interested behavior is through coercive limits on how resource systems can be used. However, subsequent scholarship has demonstrated numerous ways the tragedy of the commons can be averted through cooperative and norm-based arrangements (e.g., Ostrom 1990; National Research Council 2002). Although some discussions of climate change characterize the problem in a manner that is analogous to Hardin's model of the tragedy of the commons, this book challenges that oversimplified framing. A key question explored throughout the volume is whether the non-coercive mechanisms that allow many communities to avoid overusing their resources are likely to be successful in reining in climate change – and whether these possibilities imply a moral obligation for individuals to rein in their impacts on the climate in the absence of more formal collective efforts to address this problem.

Virtue Signaling: Engaging in conspicuous behaviors or discourse aimed at demonstrating one's moral goodness. Typically, the term is used to refer to actions and statements focused more on communicating the agent's virtuousness than on helping others, advancing justice, or making the world better.

"Voting Green":   This term refers to political support for environmentally oriented candidates and public policy measures, especially through the act of voting in elections. Although some countries have specific Green political parties, this book uses the term "voting green" in a broader sense that is compatible with supporting pro-environment candidates and policies that do not belong to such parties.

Bibliography

Achen, Christopher H., and Larry M. Bartels. 2016. *Democracy for Realists: Why Elections Do Not Produce Responsive Government.* Princeton, NJ: Princeton University Press.

Agarwal, Anil, and Sunita Narain. 1991. *Global Warming in an Unequal World: A Case of Environmental Colonialism.* New Delhi: Centre for Science and Environment.

Andreou, Chrisoula. 2010. "A Shallow Route to Environmentally-Friendly Happiness: Why Evidence that We Are Shallow Materialists Need Not Be Bad News for the Environment." *Ethics, Place & Environment* 13(1): 1–10.

Appunn, Kerstine. 2021. "The History behind Germany's Nuclear Phase-Out." *Clean Energy Wire*, March 9. www.cleanenergywire.org/factsheets/history-behind-germanys-nuclear-phase-out.

Aristotle. [~350 B.C.] 2000. *Nicomachean Ethics*, edited by Roger Crisp. New York: Cambridge University Press.

Audi, Robert, and Patrick Murphy. 2006. "The Many Faces of Integrity." *Business Ethics Quarterly* 16(1): 3–21.

Axsen, Jonn, and Michael Wolinetz. 2023. "What Does a Low-Carbon Fuel Standard Contribute to a Policy Mix? An Interdisciplinary Review of Evidence and Research Gaps." *Transport Policy* 133: 54–63.

Baatz, Christian. 2014. "Climate Change and Individual Duties to Reduce GHG Emissions." *Ethics, Policy & Environment* 17(1): 1–19.

Bacigalupi, Paolo. 2019. "A Full Life: A Science Fiction Story about America in the Age of Climate Change." *MIT Technology Review*, April 24. www.technologyreview.com/2019/04/24/135741/a-full-life/.

Barron, Alexander R., Rachel C. Venator, Ella V. H. Carlson, Jane K. Andrews, Junwen Ding, and David DeSwert. 2023. "Fossil Fuel Divestment in US Higher Education: Endowment Dependence and Temporal Dynamics." *Elementa: Science of the Anthropocene* 11(1): 00059. https://doi.org/10.1525/elementa.2023.00059.

Beavan, Colin. 2009. *No Impact Man: Adventures of a Guilty Liberal Who Attempts to Save the Planet, and the Discoveries He Makes about Himself and Our Way of Life in the Process.* New York: Macmillan.

Bernstein, Steven, and Matthew Hoffmann. 2018. "The Politics of Decarbonization and the Catalytic Impact of Subnational Climate Experiments." *Policy Sciences* 51: 189–211.

Bicchieri, Cristina. 2006. *The Grammar of Society: The Nature and Dynamics of Social Norms.* New York: Cambridge University Press.

Bonjour, S., H. Adair-Rohani, J. Wolf, N.G. Bruce, S. Mehta, A. Prüss-Ustün, et al. 2013. "Solid Fuel Use for Household Cooking: Country and Regional Estimates for 1980." *Environmental Health Perspectives* 121 (7): 784–790. https://ehp.niehs.nih.gov/doi/10.1289/ehp.1205987.

Bowles, Samuel. 2008. "Policies Designed for Self-Interested Citizens May Undermine 'The Moral Sentiments': Evidence from Economic Experiments." *Science* 320(5883): 1605–1609.

Brennan, Geoffrey. 2009. "Climate Change: A Rational Choice Politics View." *Australian Journal of Agricultural and Resource Economics* 53(3): 309–326.

Brennan, Jason. 2012. "For-Profit Business as Civic Virtue." *Journal of Business Ethics* 106(3): 313–324.

Brennan, Jason. 2016. *Against Democracy.* Princeton, NJ: Princeton University Press.

Brick, Cameron, David K. Sherman, and Heejung S. Kim. 2017. "'Green to be Seen' and 'Brown to Keep Down': Visibility Moderates the Effect of Identity on Pro-Environmental Behavior." *Journal of Environmental Psychology* 51: 226–238.

Broome, John. 2012. *Climate Matters: Ethics in a Warming World.* New York: W. W. Norton & Co.

Brownstein, Michael, Daniel Kelly, and Alex Madva. 2022. "Individualism, Structuralism, and Climate Change." *Environmental Communication* 16(2): 269–288.

Buehler, Ralph. 2011. "Determinants of Transport Mode Choice: A Comparison of Germany and the USA." *Journal of Transport Geography* 19(4): 644–657.

Cahoone, Lawrence. 2009. "Hunting as a Moral Good." *Environmental Values* 18(1): 67–89.

Caplan, Bryan. 2007. *The Myth of the Rational Voter: Why Democracies Choose Bad Policies.* Princeton, NJ: Princeton University Press.

Carter, Robert, and Erin McCarthy. 2019. "Watsuji Tetsurō." In *The Stanford Encyclopedia of Philosophy* (Winter 2019 Edition), edited by Edward N. Zalta. https://plato.stanford.edu/archives/win2019/entries/watsuji-tetsuro/.

Chislenko, Eugene. 2022. "The Role of Philosophers in Climate Change." *Journal of the American Philosophical Association* 8(4): 780–798.

Clark, Pilita. 2015. "COP21: China Accused of Blocking Progress at Paris Climate Talks." *Financial Times*, December 8. www.ft.com/content/15be0e10-9dca-11e5-b45d-4812f209f861.

Cloud, Doug. 2016. "Communicating Climate Change to Religious and Conservative Audiences: The Case of Katherine Hayhoe and Andrew Farley." *Reflections* 16(1): 57–74.

Cohen, Joshua. 1989. "Deliberation and Democratic Legitimacy." In *The Good Polity*, edited by Alan Hamlin and Philip Pettit, 17–34. New York: Basil Blackwell.

Constantino, Sara M., Gregg Sparkman, Gordon T. Kraft-Todd, Cristina Bicchieri, Damon Centola, Bettina Shell-Duncan, Sonja Vogt, and Elke U. Weber. 2022. "Scaling Up Change: A Critical Review and Practical Guide to Harnessing Social Norms for Climate Action." *Psychological Science in the Public Interest* 23(2): 50–97.

Cook, Elizabeth M., Sharon J. Hall, and Kelli L. Larson. 2012. "Residential Landscapes as Social-Ecological Systems: A Synthesis of Multi-Scalar Interactions between People and Their Home Environment." *Urban Ecosystems* 15: 19–52. https://doi.org/10.1007/s11252-011-0197-0.

Cottingham, John. 2010. "Integrity and Fragmentation." *Journal of Applied Philosophy* 27(1): 2–14.

Cox, Susan Jane Buck. 1985. "No Tragedy on the Commons." *Environmental Ethics* 7(1): 49–61.

Coyne, Christopher J. 2013. *Doing Bad by Doing Good: Why Humanitarian Action Fails*. Stanford, CA: Stanford University Press.

Cripps, Elizabeth. 2013. *Climate Change and the Moral Agent: Individual Duties in an Interdependent World*. Oxford: Oxford University Press.

Crisp, Roger, and Christopher Cowton. 1994. "Hypocrisy and Moral Seriousness." *American Philosophical Quarterly* 31(4): 343–349.

Dasgupta, Shouro, and Elizabeth J. Z. Robinson. 2022. "Attributing Changes in Food Insecurity to a Changing Climate." *Scientific Reports* 12(4709). https://doi.org/10.1038/s41598-022-08696-x.

Davis, Steven L. 2003. "The Least Harm Principle May Require that Humans Consume a Diet Containing Large Herbivores, Not a Vegan Diet." *Journal of Agricultural and Environmental Ethics* 16: 387–394.

Earthday.org. 2022. "How to Act on Climate Change: A User's Guide." www.earthday.org/campaign/act-on-climate-change/.

Easterly, William. 2006. *The White Man's Burden: Why the West's Efforts to Aid the Rest Have Done So Much Ill and So Little Good*. New York: Penguin.

Ellickson, Robert C. 1991. *Order without Law: How Neighbors Settle Disputes*. Cambridge, MA: Harvard University Press.

Environmental Defense Fund. 2021. "The True Cost of Carbon Pollution." www.edf.org/true-cost-carbon-pollution.

Fernandez, Yvette. 2022. "A Closer Look: How Water Authorities Patrol Las Vegas for Water Wasters." *Nevada Public Radio (KNPR)*, September 15. https://knpr.org/knpr/2022-09-15/a-closer-look-how-water-authorit ies-patrol-las-vegas-for-water-wasters.

Flavelle, Christopher. 2023. "As Climate Shocks Grow, Lawmakers Investigate Insurers Fleeing Risky Areas." *New York Times*, November 1. www.nytimes.com/2023/11/01/climate/climate-insurance-disasters-senate.html.

Flugum, Ryan, and Matthew E. Souther. In press. "Stakeholder Value: A Convenient Excuse for Underperforming Managers?" *Journal of Financial and Quantitative Analysis*.

Freiman, Christopher. 2021. *Why It's OK to Ignore Politics*. New York: Routledge.

Friedlingstein, P., M. O'Sullivan, M. W. Jones, et al. 2023. "Global Carbon Budget 2023." *Earth System Science Data* 15(12): 5301–5369. https://doi.org/10.5194/essd-15-5301-2023.

Friedman, Milton. 1970. "The Social Responsibility of Business Is to Increase Its Profits." *New York Times*, September 17: SM17.

Friends of the Earth. 2018. "Ending Dangerous Nuclear Power." September 24. https://foe.org/impact-stories/impact-story-3/.

Frum, David. 2021. "The West's Nuclear Mistake." *The Atlantic*, December 8. www.theatlantic.com/ideas/archive/2021/12/germany-cal ifornia-nuclear-power-climate/620888/.

Galvin, Richard, and John R. Harris. 2023. "Causal Impotence and Complicity." *Public Affairs Quarterly* 37(1): 47–63.

Gambrel, Joshua Colt, and Philip Cafaro. 2010. "The Virtue of Simplicity." *Journal of Agricultural and Environmental Ethics* 23(1–2): 85–108.

Gardiner, Stephen M. 2006. "A Perfect Moral Storm: Climate Change, Intergenerational Ethics and the Problem of Moral Corruption." *Environmental Values* 15(3): 397–413.

Gardiner, Stephen M. 2010. "Is 'Arming the Future' with Geoengineering Really the Lesser Evil? Some Doubts About the Ethics of Intentionally Manipulating the Climate System." In *Climate Ethics: Essential Readings*, edited by Stephen M. Gardiner, Simon Caney, Dale Jamieson, and Henry Shue, 284–312. New York: Oxford University Press.

Gardiner, Stephen M. 2011. *A Perfect Moral Storm: The Ethical Tragedy of Climate Change*. New York: Oxford University Press.

Gardiner, Stephen M. 2022. "Is the Paris Climate Agreement Another Dangerous Illusion?" *Norwegian Academy of Science and Letters Yearbook* 2021: 157–174.

Gardiner, Stephen M., and Arthur Obst. 2023. *Dialogues on Climate Justice*. New York: Routledge.

Gates, Jimmie E. 2018. "New Law Will Prohibit Local Government from Banning Plastic Bags or Other Type Containers." *Mississippi Clarion*

Ledger, March 20. www.clarionledger.com/story/news/politics/2018/
03/20/new-law-prohibit-local-government-banning-plastic-bags-other-
type-containers/442256002/.

Gaus, Gerald F. 2008. *On Philosophy, Politics, and Economics*. Belmont,
CA: Thomson Wadsworth.

Gaus, Gerald F., and John Thrasher. 2021. *Philosophy, Politics, and
Economics: An Introduction*. Princeton, NJ: Princeton University Press.

Gertler, Aaron. 2021. "The Effective Altruism Handbook." https://forum.
effectivealtruism.org/handbook.

Global Witness. 2022. *Decade of Defiance: Ten Years of Reporting Land
and Environmental Activism Worldwide*. www.globalwitness.org/doc-
uments/20425/Decade_of_defiance_EN_-_September_2022.pdf.

Goddard, Mark A., Andrew J. Dougill, and Tim G. Benton. 2013. "Why
Garden for Wildlife? Social and Ecological Drivers, Motivations and
Barriers for Biodiversity Management in Residential Landscapes."
Ecological Economics 86: 258–273.

Godoy, Eric S. 2017. "Sharing Responsibility for Divesting from Fossil
Fuels." *Environmental Values* 26(6): 693–710.

Goldstein, N. J., Robert B. Cialdini, and Vladas Griskevicius. 2008. "A
Room with a Viewpoint: Using Social Norms to Motivate Environmental
Conservation in Hotels." *Journal of Consumer Research* 35: 472–482.

Göransson, Jessica, and Henrik Andersson. 2023. "Factors That Make
Public Transport Systems Attractive: A Review of Travel Preferences
and Travel Mode Choices." *European Transportation Research Review*
15(32). https://doi-org.coloradocollege.idm.oclc.org/10.1186/s12
544-023-00609-x.

Gramling, Carolyn. 2023. "COP28 Nations Agreed to 'Transition' from
Fossil Fuels. That's Too Slow, Experts Say." *ScienceNews*, December 15.
www.sciencenews.org/article/cop28-fossil-fuels-climate-change.

Grantham Institute. 2019. "9 Things You Can Do about Climate Change."
www.imperial.ac.uk/stories/climate-action/.

Graton, Aurélien, François Ric, and Emilie Gonzalez. 2016. "Reparation
or Reactance? The Influence of Guilt on Reaction to Persuasive
Communication." *Journal of Experimental Social Psychology* 62: 40–49.

Greenfield, Patrick. 2022. "More Than 1,700 Environmental Activists
Murdered in the Past Decade – Report." *The Guardian*, September 28.
www.theguardian.com/environment/2022/sep/29/global-witness-report-
1700-activists-murdered-past-decade-aoe.

Greenpeace. 2022. "Nuclear Energy." www.greenpeace.org/usa/fighting-
climate-chaos/issues/nuclear/.

Griskevicius, Vladas, Joshua M. Tybur, and Bram Van den Bergh. 2010.
"Going Green to Be Seen: Status, Reputation, and Conspicuous
Conservation." *Journal of Personality and Social Psychology*
98(3): 392–404.

Gromet, Dena M., Howard Kunreuther, and Richard P. Larrick. 2013. "Political Ideology Affects Energy-Efficiency Attitudes and Choices." *PNAS* 110(23): 9314–9319.

Haidt, Jonathan. 2012. *The Righteous Mind.* New York: Random House.

Hakli, Raul, Kaarlo Miller, and Raimo Tuomela. 2010. "Two Kinds of We-Reasoning." *Economics & Philosophy* 26(3): 291–320.

Halcoussis, Dennis, and Anton D. Lowenberg. 2019. "The Effects of the Fossil Fuel Divestment Campaign on Stock Returns." *The North American Journal of Economics and Finance* 47: 669–674.

Hardin, Garrett. 1968. "The Tragedy of the Commons." *Science* 162(3859): 1243–1248.

Harrison, Jonathan. 1979. "Rule Utilitarianism and Cumulative-Effect Utilitarianism." *Canadian Journal of Philosophy Supplementary Volume* 5: 21–45.

Hart, P. Sol, and Erik C. Nisbet. 2012. "Boomerang Effects in Science Communication: How Motivated Reasoning and Identity Cues Amplify Opinion Polarization About Climate Mitigation Policies." *Communication Research* 39(6): 701–723.

Hartzell, Lauren. 2011. "Responsibility for Emissions: A Commentary on John Nolt's 'How Harmful are the Average American's Greenhouse Gas Emissions?'" *Ethics, Policy & Environment* 14(1): 15–17.

Hayek, F.A. 1945. "The Use of Knowledge in Society." *American Economic Review* 35(4): 519–530.

Hayter, Julian Maxwell. 2023. "Civil Rights Legislation Sparked Powerful Backlash That's Still Shaping American Politics." *The Conversation*, April 3. https://theconversation.com/civil-rights-legislation-sparked-powerful-backlash-thats-still-shaping-american-politics-197475.

Heath, Joseph. 2021. *The Philosophical Foundations of Climate Change Policy.* New York: Oxford University Press.

Hedberg, Trevor. 2018. "Climate Change, Moral Integrity, and Obligations to Reduce Individual Greenhouse Gas Emissions." *Ethics, Policy & Environment* 21(1): 64–80.

Heglar, Mary Annaise. 2019. "I Work in the Environmental Movement. I Don't Care If You Recycle." *Vox*, June 4. www.vox.com/the-highlight/2019/5/28/18629833/climate-change-2019-green-new-deal.

Hiller, Avram. 2011. "Climate Change and Individual Responsibility." *The Monist* 94(3): 349–368.

Hobbes, Thomas. 1968. *Leviathan.* New York: Penguin.

Hornsey, Matthew J., and Kelly S. Fielding. 2020. "Understanding (and Reducing) Inaction on Climate Change." *Social Issues and Policy Review* 14(1): 3–35.

Hourdequin, Marion. 2010. "Climate, Collective Action and Individual Ethical Obligations." *Environmental Values* 19(4): 443–464.

Hourdequin, Marion. 2011. "Climate Change and Individual Responsibility: A Reply to Johnson." *Environmental Values* 20(2): 157–162.

Hourdequin, Marion. 2015. *Environmental Ethics: From Theory to Practice*. London: Bloomsbury.

Hourdequin, Marion. 2021. "Environmental Ethics: The State of the Question." *The Southern Journal of Philosophy* 59(3): 270–308.

Hourdequin, Marion, and David B. Wong. 2005. "A Relational Approach to Environmental Ethics." *Journal of Chinese Philosophy* 32(1): 19–33.

Hultman, Nathan E., Leon Clarke, Carla Frisch, Kevin Kennedy, Haewon McJeon, Tom Cyrs, Pete Hansel et al. 2020. "Fusing Subnational with National Climate Action Is Central to Decarbonization: The Case of the United States." *Nature Communications* 11(1): 5255.

Hurst, Kristin F., and Nicole D. Sintov. 2022. "Guilt Consistently Motivates Pro-environmental Outcomes while Pride Depends on Context." *Journal of Environmental Psychology* 80: 101776. https://doi.org/10.1016/j.jenvp.2022.101776.

IEA (International Energy Agency). 2019. *Nuclear Power in a Clean Energy System*. Paris: International Energy Agency.

IPCC (Intergovernmental Panel on Climate Change). 1996. *Climate Change 1995: Economic and Social Dimensions of Climate Change, Contribution of Working Group III to the Second Assessment Report of the Intergovernmental Panel on Climate Change*. New York: Cambridge University Press.

IPCC. 2018a. "Summary for Policymakers." In *Global Warming of 1.5°C. An IPCC Special Report on the Impacts of Global Warming of 1.5°C above Pre-Industrial Levels and Related Global Greenhouse Gas Emission Pathways, in the Context of Strengthening the Global Response to the Threat of Climate Change, Sustainable Development, and Efforts to Eradicate Poverty*, edited by V. Masson-Delmotte et al., 3–24. New York: Cambridge University Press. https://doi.org/10.1017/9781009157940.001.

IPCC. 2018b. *Global Warming of 1.5°C: An IPCC Special Report on the Impacts of Global Warming of 1.5°C above Pre-Industrial Levels and Related Global Greenhouse Gas Emission Pathways, in the Context of Strengthening the Global Response to the Threat of Climate Change, Sustainable Development, and Efforts to Eradicate Poverty*, edited by V. Masson-Delmotte et al. New York: Cambridge University Press. https://doi.org/10.1017/9781009157940.

IPCC. 2021. *Climate Change 2021: The Physical Science Basis – Working Group I Contribution to the Sixth Assessment Report*. New York: Cambridge University Press.

IPCC. 2022a. *Climate Change 2022: Impacts, Adaptation, and Vulnerability – Working Group II Contribution to the Sixth Assessment Report.* New York: Cambridge University Press.

IPCC. 2022b. *Climate Change 2022: Mitigation of Climate Change – Working Group III Contribution to the Sixth Assessment Report.* New York: Cambridge University Press.

Jamieson, Dale. 1992. "Ethics, Public Policy, and Global Warming." *Science, Technology, and Human Values* 17(2): 139–153.

Jamieson, Dale. 2007. "When Utilitarians Should Be Virtue Theorists." *Utilitas* 19(2): 160–183.

Jamieson, Dale. 2010. "Climate Change, Responsibility, and Justice." *Science and Engineering Ethics* 16: 441–445.

Johnson, Baylor. 2003. "Ethical Obligations in a Tragedy of the Commons." *Environmental Values* 12(3): 271–287.

Johnson, Baylor. 2011. "The Possibility of a Joint Communique: My Response to Hourdequin." *Environmental Values* 20(2): 147–156.

Kahan, Dan M. 2010. "Fixing the Communications Failure." *Nature* 463(7279): 296–297.

Kahan, Dan M. 2012. "Why We Are Poles Apart on Climate Change." *Nature* 488: 255.

Kahan, Dan M. 2013. "Ideology, Motivated Reasoning, and Cognitive Reflection." *Judgment and Decision Making* 8(4): 407–424.

Kahneman, Daniel. 2011. *Thinking Fast and Slow.* New York: Farrar, Straus and Giroux.

Kant, Immanuel. [1785] 2006. *Groundwork of the Metaphysics of Morals*, translated and edited by Mary Gregor. New York: Cambridge University Press.

Kant, Immanuel. [1797] 1996. "The Metaphysics of Morals." In *The Cambridge Edition of the Works of Immanuel Kant: Practical Philosophy*, edited by Mary J. Gregor, 353–603. New York: Cambridge University Press.

Kawall, Jason. 2011. "Future Harms and Current Offspring." *Ethics, Policy and Environment* 14(1): 23–26.

Khalfan, Ashfaq, Astrid Nilsson Lewis, Carlos Aguilar, Jacqueline Persson, Max Lawson, Nafkote Dabi, Safa Jayoussi, Sunil AcharyaAstrid Nilsson Lewis, Carlos Aguilar, Jacqueline Persson, Max Lawson, Nafkote Dabi, Safa Jayoussi, and Sunil Acharya. 2023. *Climate Equality: A Planet for the 99%.* Oxford: Oxfam. https://policy-practice.oxfam.org/resources/climate-equality-a-planet-for-the-99-621551/.

Kingston, Ewan, and Walter Sinnott-Armstrong. 2018. "What's Wrong with Joyguzzling?" *Ethical Theory and Moral Practice* 21(1): 169–186.

Klein, Naomi. 2015. *This Changes Everything: Capitalism vs. the Climate.* New York: Simon and Schuster.

Koyama, Mark, and Jared Rubin. 2022. *How the World Became Rich: The Historical Origins of Economic Growth*. Medford, MA: Polity Press.

Kumar, Deepak, Gautam Kalghatgi, and Avinash Kumar Agarwal. 2023. "Comparison of Economic Viability of Electric and Internal Combustion Engine Vehicles Based on Total Cost of Ownership Analysis." In *Transportation Systems Technology and Integrated Management*, edited by R.K. Upadhyay, S.K Sharma, V. Kumar, and H. Valera, 455–489. Springer. https://doi.org/10.1007/978-981-99-1517-0_20.

Kutz, Christopher. 2000. *Complicity: Ethics and Law for a Collective Age*. New York: Cambridge University Press.

Lawson, Brian. 2013. "Individual Complicity in Collective Wrongdoing." *Ethical Theory and Moral Practice* 16: 227–243.

Leiserowitz, A., E. Maibach, S. Rosenthal, J. Kotcher, L. Neyens, J. Carman, J. Marlon, K. Lacroix, and M. Goldberg. 2021a. *Consumer Activism on Global Warming, September 2021*. New Haven, CT: Yale Program on Climate Change Communication.

Leiserowitz, A., E. Maibach, S. Rosenthal, J. Kotcher, J. Carman, L. Neyens, M. Goldberg, K. Lacroix, and J. Marlon. 2021b. *Politics and Global Warming, September 2021*. New Haven, CT: Yale Program on Climate Change Communication.

Liu, Zhe, Juhyun Song, Joseph Kubal, Naresh Susarla, Kevin W. Knehr, Ehsan Islam, Paul Nelson, and Shabbir Ahmed. 2021. "Comparing Total Cost of Ownership of Battery Electric Vehicles and Internal Combustion Engine Vehicles." *Energy Policy* 158: 112564. https://doi.org/10.1016/j.enpol.2021.112564

Lofthouse, Jordan K., and Roberta Q. Herzberg. 2023. "The Continuing Case for a Polycentric Approach to Coping with Climate Change." *Sustainability* 15(4): art. 3770.

Lomborg, Bjørn. 2008. *Cool It: The Skeptical Environmentalist's Guide to Global Warming*. New York: Vintage Books.

Lorenz, Edward N. 1972. "Predictability: Does the Flap of a Butterfly's Wings in Brazil Set Off a Tornado in Texas?" Presentation to the 139th meeting of the American Association for the Advancement of Science, in Washington, DC, December 29, 1972.

Lovins, Amory B., and M.V. Ramana. 2021. "Three Myths about Renewable Energy and the Grid, Debunked." *Yale Environment 360*, December 9. https://e360.yale.edu/features/three-myths-about-renewable-energy-and-the-grid-debunked.

Mac, Ryan. 2023. "Allstate Is No Longer Offering New Policies in California." *New York Times*, June 4. www.nytimes.com/2023/06/04/business/allstate-insurance-california.html.

MacAskill, William. 2015. *Doing Good Better: How Effective Altruism Can Help You Make a Difference*. New York: Avery.

MacLean, Douglas. 2019. "Climate Complicity and Individual Accountability." *The Monist* 102(1): 1–21.

Malima, Gabriel Clement, and Francis Moyo. 2023. "Are Electric Vehicles Economically Viable in Sub-Saharan Africa? The Total Cost of Ownership of Internal Combustion Engine and Electric Vehicles in Tanzania." *Transport Policy* 141: 14–26.

Maltais, Aaron. 2013. "Radically Non-Ideal Climate Politics and the Obligation to at Least Vote Green." *Environmental Values* 22(5): 589–608.

Mankiw, N. Gregory. 2021. *Principles of Economics*, 9th ed. Boston, MA: Cengage.

Markowitz, Ezra M., and Azim F. Shariff. 2012. "Climate Change and Moral Judgement." *Nature Climate Change* 2(4): 243–247.

Marszal, Andrew. 2023. "The 'Water Cops' of Las Vegas Make City a Model in Drought-Hit US." May 8. https://phys.org/news/2023-05-cops-las-vegas-city-drought-hit.html.

Marx, Karl. [1846] 2000. "The German Ideology." In *Karl Marx: Selected Writings*, edited by David McLellan. New York: Oxford University Press.

Mayer, Seth. 2019. "Climate Change and Environmental Regulation." In *Ethics, Left and Right: The Moral Issues That Divide Us*, edited by Bob Fischer, 285–294. New York: Oxford University Press.

McGregor, Deborah. 2009. "Honouring Our Relations: An Anishnaabe Perspective on Environmental Justice." In *Speaking for Ourselves: Environmental Justice in Canada*, edited by J. Agyeman, P. Cole, R. Haluza DeLay, and P. O'Riley, 27–41. Vancouver: UBC Press.

McKibben, Bill. 2015. "Falling Short on Climate Change." *The New York Times*, December 14, A23.

Metz, Thaddeus, and Sarah Clark Miller. 2016. "Relational Ethics." In *The International Encyclopedia of Ethics*, edited by H. LaFollette. John Wiley & Sons. https://doi.org/10.1002/9781444367072.wbiee826.

Mill, John Stuart. 1844. "On the Definition of Political Economy; and On the Method of Investigation Proper to It." In *Essays on Some Unsettled Questions of Political Economy*, 120–164. London: John W. Parker.

Monbiot, George. 2007. "Ethical Shopping Is Just Another Way of Showing How Rich You Are." *The Guardian*, July 23. www.theguardian.com/commentisfree/2007/jul/24/comment.businesscomment.

Monserrate, Steven Gonzalez. 2022. "The Cloud Is Material: On the Environmental Impacts of Computation and Data Storage." *MIT Case Studies in Social and Ethical Responsibilities of Computing*, Winter (January). https://doi.org/10.21428/2c646de5.031d4553.

Moyo, Dambisa. 2009. *Dead Aid: Why Aid Is Not Working and How There Is a Better Way for Africa*. New York: Farrar, Straus and Giroux.

Nafkote Dabi, Safa Jayoussi, Sunil Acharya, Astrid Nilsson Lewis, Carlos Aguilar, Jacqueline Persson, Max Lawson. 2023. *Climate Equality: A*

Planet for the 99%. Oxford: Oxfam. https://policy-practice.oxfam.org/resources/climate-equality-a-planet-for-the-99-621551/.

Nassauer, Joan Iverson, Zhifang Wang, and Erik Dayrell. 2009. "What Will the Neighbors Think? Cultural Norms and Ecological Design." *Landscape and Urban Planning* 92(3–4): 282–292.

National Research Council. 2002. *The Drama of the Commons*. Washington, DC: National Academy Press.

Natural Resources Defense Council. 2022. "How You Can Stop Global Warming." www.nrdc.org/stories/how-you-can-stop-global-warming.

Nielsen, Kristian S., Kimberly A. Nicholas, Felix Creutzig, Thomas Dietz, and Paul C. Stern. 2021. "The Role of High-Socioeconomic-Status People in Locking in or Rapidly Reducing Energy-Driven Greenhouse Gas Emissions." *Nature Energy* 6(11): 1011–1016.

Nolt, John. 2011. "How Harmful Are the Average American's Greenhouse Gas Emissions?" *Ethics, Policy & Environment* 14(1): 3–10.

Nordhaus, William. 2013. *The Climate Casino: Risk, Uncertainty, and Economics for a Warming World*. New Haven, CT: Yale University Press.

Norgaard, Kari Marie. 2011. *Living in Denial: Climate Change, Emotions, and Everyday Life*. Cambridge, MA: MIT Press.

Odenbaugh, Jay. 2011. "This American Life." *Ethics, Policy & Environment* 14(1): 27–29.

Oreskes, Naomi, and Erik M. Conway. 2011. *Merchants of Doubt: How a Handful of Scientists Obscured the Truth on Issues from Tobacco Smoke to Global Warming*. Bloomsbury Publishing.

Ortmann, Jakob, and Walter Veit. 2023. "Theory Roulette: Choosing that Climate Change is not a Tragedy of the Commons." *Environmental Values* 32(1): 65–89.

Ostrom, Elinor. 1990. *Governing the Commons: The Evolution of Institutions for Collective Action*. New York: Cambridge University Press.

Ostrom, Elinor. 2009a. "A Polycentric Approach for Coping with Climate Change." Background Paper to the 2010 World Development Report, Policy Research Working Paper #5095. Washington, DC: World Bank.

Ostrom, Elinor. 2009b. "Building Trust to Solve Commons Dilemmas: Taking Small Steps to Test an Evolving Theory of Collective Action." In *Games, Groups, and the Global Good*, edited by S. A. Levin, 17–35. Berlin: Springer. https://doi-org.coloradocollege.idm.oclc.org/10.1007/978-3-540-85436-4_13.

Ostrom, Elinor. 2010. "Polycentric Systems for Coping with Collective Action and Global Environmental Change." *Global Environmental Change* 20(4): 550–557.

Ostrom, Elinor. 2012. "Nested Externalities and Polycentric Institutions: Must We Wait for Global Solutions to Climate Change Before Taking Action at Other Scales?" *Economic Theory* 49(2): 353–369.

Ostrom, Elinor, Marco A. Janssen, and John M. Anderies. 2007. "Going Beyond Panaceas." *Proceedings of the National Academy of Sciences* 104(39): 15176–15178.

Palanski, Michael E., Surinder S. Kahai, and Francis J. Yammarino. 2011. "Team Virtues and Performance: An Examination of Transparency, Behavioral Integrity, and Trust." *Journal of Business Ethics* 99: 201–216.

Patt, Anthony. 2017. "Beyond the Tragedy of the Commons: Reframing Effective Climate Change Governance." *Energy Research & Social Science* 34: 1–3.

Pearce, Fred. 2024. "Mind the Gaps: How the UN Climate Plan Fails to Follow the Science." *Yale360*, January 9. https://e360.yale.edu/features/un-climate-science-1.5-net-zero.

Plantinga, Auke, and Bert Scholtens. 2021. "The Financial Impact of Fossil Fuel Divestment." *Climate Policy* 21(1): 107–119.

Portmore, Douglas W. 2012. "Imperfect Reasons and Rational Options." *Noûs* 46(1): 24–60.

Prinzing, Michael. 2023. "Going Green Is Good for You: Why We Need to Change the Way We Think about Pro-Environmental Behavior." *Ethics, Policy & Environment* 26(1): 1–18.

Radzvilas, Mantas, and Jurgis Karpus. 2021. "Team Reasoning Without a Hive Mind." *Research in Economics* 75(4): 345–353.

Rapson, David S., and Erich Muehlegger. 2023. "The Economics of Electric Vehicles." *Review of Environmental Economics and Policy* 17(2): 274–294.

Raterman, Ty. 2012. "Bearing the Weight of the World: On the Extent of an Individual's Environmental Responsibility." *Environmental Values* 21(4): 417–436.

Raymond, Leigh, Daniel Kelly, and Erin P. Hennes. In press. "Norm-based Governance for Severe Collective Action Problems: Lessons from Climate Change and COVID-19." *Perspectives on Politics*.

Rees, Jonas H., Sabine Klug, and Sebastian Bamberg. 2015. "Guilty Conscience: Motivating Pro-environmental Behavior by Inducing Negative Moral Emotions." *Climatic Change* 130: 439–452.

Ritchie, Hannah, and Max Roser. 2020. "CO_2 and Greenhouse Gas Emissions." *Our World in Data*. https://ourworldindata.org/co2-and-other-greenhouse-gas-emissions.

Ritchie, Hannah, Pablo Rosado, and Max Roser. 2023. "CO_2 and Greenhouse Gas Emissions." Published online at OurWorldInData.org. https://ourworldindata.org/co2-and-greenhouse-gas-emissions. [Note: Data derived in part from Friedlingstein et al. 2023.]

Robbins, Lionel. 1932. *An Essay on the Nature and Significance of Economic Science*. London: Macmillan.

Robbins, Paul. 2007. *Lawn People: How Grasses, Weeds, and Chemicals Make Us Who We Are*. Philadelphia, PA: Temple University Press.

Rosenthal, Elisabeth. 2008. "Motivated by a Tax, Irish Spurn Plastic Bags." *New York Times*, February 2. www.nytimes.com/2008/02/02/world/europe/02bags.html.

Roser, Max, and Esteban Ortiz-Ospina. 2019. "Global Extreme Poverty." *Our World in Data*. https://ourworldindata.org/extreme-poverty.

Roser, Max, Esteban Ortiz-Ospina, and Hannah Ritchie. 2019. "Life Expectancy." *Our World in Data*. https://ourworldindata.org/life-expectancy.

Roser, Max, Hannah Ritchie, and Bernadeta Dadonaite. 2019. "Child and Infant Mortality." *Our World in Data*. https://ourworldindata.org/child-mortality.

Safford, Hannah, Elizabeth Larry, E. Gregory McPherson, David J. Nowak, and Lynn M. Westphal. 2013. "Urban Forests and Climate Change." www.climatehubs.usda.gov/sites/default/files/Urban-Forests_CCRC.pdf

Sandberg, Joakim. 2011. "'My Emissions Make No Difference': Climate Change and the Argument from Inconsequentialism." *Environmental Ethics* 33(3): 229–248.

Sandler, Ronald. 2010. "Ethical Theory and the Problem of Inconsequentialism: Why Environmental Ethicists Should Be Virtue-Oriented Theorists." *Journal of Agricultural and Environmental Ethics* 23(1–2): 167–183.

Sandler, Ronald. 2011. "Beware of Averages: A Response to John Nolt's 'How Harmful are the Average American's Greenhouse Gas Emissions?'" *Ethics, Policy & Environment* 14(1): 31–33.

Sandler, Ronald, and Philip Cafaro. 2005. *Environmental Virtue Ethics*. Lanham, MD: Rowman & Littlefield.

Saunders, Leland F. 2021. "Virtues as Reasons Structures." *Philosophical Studies* 178 (9): 2785–2804.

Schmidtz, David. 2016. "After Solipsism." *Oxford Studies in Normative Ethics* 6: 145–165.

Schmidtz, David, and Elizabeth Willott. 2006. "Varieties of Overconsumption." *Ethics, Place & Environment* 9(3): 351–365.

Schwenkenbecher, Anne. 2014. "Is There an Obligation to Reduce One's Individual Carbon Footprint?" *Critical Review of International Social and Political Philosophy* 17(2): 168–188.

Schwenkenbecher, Anne. 2019. "Collective Moral Obligations: 'We-Reasoning' and the Perspective of the Deliberating Agent." *The Monist* 102(2): 151–171.

Schwenkenbecher, Anne. 2021. *Getting Our Act Together: A Theory of Collective Moral Obligations*. New York: Routledge.

Sen, Amartya. 1981. *Poverty and Famines: An Essay on Entitlement and Deprivation*. New York: Oxford University Press.

Sexton, Steven E., and Alison L. Sexton. 2014. "Conspicuous Conservation: The Prius Halo and Willingness to Pay for Environmental

Bona Fides." *Journal of Environmental Economics and Management* 67(3): 303–317.

Shahar, Dan C. 2016. "Treading Lightly on the Climate in a Problem-Ridden World." *Ethics, Policy & Environment* 19(2): 183–195.

Shahar, Dan C. 2019. "Sustaining Growth." In *Ethics, Left and Right: The Moral Issues that Divide Us*, edited by Bob Fischer, 294–301. New York: Oxford University Press.

Shahar, Dan C. 2022. *Why It's OK to Eat Meat*. New York: Routledge.

Shahar, Dan C. 2024. "Human Flourishing and Our Relationships with Nature." *Ethics & the Environment* 29(1): 89–108.

Shellenberger, Michael. 2016. "How Fear of Nuclear Power is Hurting the Environment." www.ted.com/talks/michael_shellenberger_how_fear_of_nuclear_power_is_hurting_the_environment.

Shue, Henry. 1993. "Subsistence Emissions and Luxury Emissions." *Law and Policy* 15(1): 39–60.

Singer, Peter. 2009. *The Life You Can Save: Acting Now to End World Poverty*. New York: Random House.

Singh, M., S. Tuli, R.J. Butcher, R. Kaur, and S.S. Gill. 2021. "Dynamic Shift from Cloud Computing to Industry 4.0: Eco-Friendly Choice or Climate Change Threat?" In *IoT-based Intelligent Modelling for Environmental and Ecological Engineering* (Lecture Notes on Data Engineering and Communications Technologies, vol 67), edited by P. Krause and F. Xhafa, 275–293. Springer. https://doi.org/10.1007/978-3-030-71172-6_12.

Sinnott-Armstrong, Walter. 2005. "It's Not My Fault: Global Warming and Individual Moral Obligations." In *Perspectives on Climate Change: Science, Politics, Ethics*, edited by Walter Sinnott-Armstrong and Richard B. Howarth, 285–307. Amsterdam: Elsevier.

Smiley, Marion. 2023. "Collective Responsibility." In *The Stanford Encyclopedia of Philosophy* (Fall 2023 Edition), edited by E. N. Zalta and U. Nodelman. https://plato.stanford.edu/archives/fall2023/entries/collective-responsibility/.

Smith, Adam. [1776] 1904. *An Inquiry into the Nature and Causes of the Wealth of Nations*. London: Methuen & Co.

Smith, Alisa, and J. B. MacKinnon. 2007. *Plenty: Eating Locally on the 100 Mile Diet*. New York: Three Rivers Press.

Somin, Ilya. 2016. *Democracy and Political Ignorance: Why Smaller Government is Smarter*. Stanford, CA: Stanford University Press.

State Farm. 2023. "State Farm General Insurance Company: California New Business Update." https://newsroom.statefarm.com/state-farm-general-insurance-company-california-new-business-update.

Stern, Nicholas H. 2006. *The Economics of Climate Change: The Stern Review*. Cambridge: Cambridge University Press.

Stern, Nicholas, Joseph Stiglitz, Kristina Karlsson, and Charlotte Taylor. 2022. "A Social Cost of Carbon Consistent with a Net-Zero Climate Goal." Roosevelt Institute. https://rooseveltinstitute.org/wp-content/uploads/2022/01/RI_Social-Cost-of-Carbon_202201-1.pdf.

Surey, Bryan, Dayananda Palihawadana, Charalampos Saridakis, and Aristeidis Theotokis. 2020. "How Downplaying Product Greenness Affects Performance Evaluations: Examining the Effects of Implicit and Explicit Green Signals in Advertising." *Journal of Advertising* 49(2): 125–140.

Swenson, Ali. 2023. "Moms for Liberty Rises as Power Player in GOP Politics after Attacking Schools over Gender, Race." *Associated Press*, June 11. https://apnews.com/article/moms-for-liberty-2024-election-republican-candidates-f46500e0e17761a7e6a3c02b61a3d229.

Tabuchi, Hiroko. 2016. "'Rolling Coal' in Diesel Trucks, to Rebel and Provoke." *The New York Times*, September 4. www.nytimes.com/2016/09/05/business/energy-environment/rolling-coal-in-diesel-trucks-to-rebel-and-provoke.html.

Tam, Agnes. 2020. "Why Moral Reasoning Is Insufficient for Moral Progress." *Journal of Political Philosophy* 28(1): 73–96.

Todd, Benjamin, and the 80,000 Hours Team. 2016. *80,000 Hours: Find a Fulfilling Career That Does Good*. Oxford: Centre for Effective Altruism.

Tol, Richard S.J. 2018. "The Economic Impacts of Climate Change." *Review of Environmental Economics and Policy*: 4–25.

Trinks, Arjan, Bert Scholtens, Machiel Mulder, and Lammertjan Dam. 2018. "Fossil Fuel Divestment and Portfolio Performance." *Ecological Economics* 146: 740–748.

UNECE (United Nations Economic Commission for Europe). 2021. *Technology Brief: Nuclear Power*. Geneva: United Nations Economic Commission for Europe.

UNFCCC (United Nations Framework Convention on Climate Change). 2021. "8 Ways You Can Take Climate Action Right Now." https://unfccc.int/blog/8-ways-you-can-take-climate-action-right-now.

UNFCCC. 2022a. "Moving Towards the Enhanced Transparency Framework." https://unfccc.int/enhanced-transparency-framework.

UNFCCC. 2022b. "Nationally Determined Contributions (NDCs)." https://unfccc.int/process-and-meetings/the-paris-agreement/nationally-determined-contributions-ndcs/.

UNFCCC. 2022c. "Global Stocktake." https://unfccc.int/topics/global-stocktake/global-stocktake.

UNFCCC. 2023. "Draft Decision: Outcome of the First Global Stocktake." FCCC/PA/CMA/2023/L.17. https://unfccc.int/sites/default/files/resource/cma2023_L17_adv.pdf.

UNHCR (United Nations Office of the High Commissioner on Human Rights). 2022. "Climate Change the Greatest Threat the World Has Ever

Faced, UN Expert Warns." October 21. www.ohchr.org/en/press-relea
ses/2022/10/climate-change-greatest-threat-world-has-ever-faced-un-
expert-warns.

United Nations. 1992. *United Nations Framework Convention on Climate Change.* United Nations. https://unfccc.int/files/essential_background/
background_publications_htmlpdf/application/pdf/conveng.pdf.

United Nations. 2015. *Paris Agreement.* https://unfccc.int/sites/default/
files/english_paris_agreement.pdf.

United Nations. 2018. "Climate Change: An 'Existential Threat' to Humanity, UN Chief Warns Global Summit." UN News. May 18.
https://news.un.org/en/story/2018/05/1009782.

United Nations. 2021. "COP26 Closes with 'Compromise' Deal on Climate, but It's 'Not Enough,' says UN Chief." UN News. November 13. https://news.un.org/en/story/2021/11/1105792.

United Nations. 2023a. "About Us." www.un.org/en/about-us.

United Nations. 2023b. "Actions for a Healthy Planet." www.un.org/en/
actnow/ten-actions.

U.S. Department of State. 2015. "COP21 Press Availability with Special Envoy Todd Stern." December 2. https://2009-2017.state.gov/s/climate/
releases/2015/250305.htm.

U.S. DOE (United States Department of Energy Office of Energy Efficiency & Renewable Energy). 2022. "Compare Side-by-Side: 2022 Toyota Tacoma 4WD vs. 2022 Toyota Prius." www.fueleconomy.gov/feg/Find.
do?action=sbs&id=44478&id=44079.

U.S. EPA (United States Environmental Protection Agency). 2022. "Climate Change Resources for Educators and Students." www.epa.gov/climate-
change/climate-change-resources-educators-and-students.

U.S. EPA. 2024. "Greenhouse Gases Equivalencies Calculator – Calculations and References." www.epa.gov/energy/greenhouse-gases-
equivalencies-calculator-calculations-and-references.

U.S. IWG (United States Government Interagency Working Group on Social Cost of Greenhouse Gases). 2021. "Technical Support Document: Social Cost of Carbon, Methane, and Nitrous Oxide: Interim Estimates under Executive Order 13990." www.whitehouse.gov/wp-content/uploads/
2021/02/TechnicalSupportDocument_SocialCostofCarbonMethaneN
itrousOxide.pdf.

Vanderheiden, Steve. 2016. "Climate Justice Beyond International Burden Sharing." *Midwest Studies in Philosophy* 40: 27–42.

Watts, Jonathan. 2019. "Greta Thunberg Sets Sail for New York on Zero-Carbon Yacht." *The Guardian*, August 14. www.theguardian.
com/environment/2019/aug/14/greta-thunberg-sets-sail-plymouth-clim
ate-us-trump.

Weber, Elke U. 2010. "What Shapes Perceptions of Climate Change?" *Wiley Interdisciplinary Reviews: Climate Change* 1(3): 332–342.

Weldon, Peter, Patrick Morrissey, and Margaret O'Mahony. 2018. "Long-Term Cost of Ownership Comparative Analysis between Electric Vehicles and Internal Combustion Engine Vehicles." *Sustainable Cities and Society* 39: 578–591.

Whiteman, Gail. 2009. "All My Relations: Understanding Perceptions of Justice and Conflict Between Companies and Indigenous Peoples." *Organization Studies* 30(1): 101–120.

Whyte, Kyle. 2020. "Too Late for Indigenous Climate Justice: Ecological and Relational Tipping Points." *Wiley Interdisciplinary Reviews: Climate Change* 11(1): e603. https://doi.org/10.1002/wcc.603.

Wicksteed, Philip. 1910. "Business and the Economic Nexus." In *The Common Sense of Political Economy: Including a Study of the Human Basis of Economic Law*. London: Macmillan & Co.

Wong, David B. 2011. "Agon and Hé: Contest and Harmony." In *Ethics in Early China: An Anthology*, edited by C. Fraser, D. Robins, and T. O'Leary, 197–216. Hong Kong: Hong Kong University Press.

Yoder, Landon, Courtney Hammond Wagner, Kira Sullivan-Wiley, and Gemma Smith. 2022. "The Promise of Collective Action for Large-Scale Commons Dilemmas: Reflections on Common-Pool-Resource Theory." *International Journal of the Commons* 16(1): 47–63. www.jstor.org/stable/48712053.

Young, Iris Marion. 2006. "Responsibility and Global Justice: A Social Connection Model." *Social Philosophy and Policy* 23(1): 102–130.

Young, Iris Marion. 2011. *Responsibility for Justice*. New York: Oxford University Press.

Zaraska, Mara. 2022. "Virtue Signaling for the Environment Isn't Working." *Breakthrough Institute Journal* 18 (Fall). https://thebreakthrough.org/journal/no-18-fall-2022/virtue-signaling-for-the-environment-isnt-working.

Index

aggregation of risks 117–120
attribution 14
Audi, Robert 22
autonomy xv, 67–70, 113, 155–158

Bicchieri, Cristina 66
Broome, John 115–121

carbon footprint 15, 18
chaotic systems 116–117, 178–179
command-and-control regulations 67
complementarity 153–154
complicity 13, 55–56, 179
consequentialism 8, 30, 115
conspicuous conservation (see virtue signaling)
costs of climate action 62–63, 67–70, 78–79, 84–85
cumulative effects 9, 141–144

"dirty hands" objection 105–107
division of labor xii–xiii, 53–54, 79, 127–128, 149–151, 155–158, 180
duties (see obligations)

economic growth xii, 47–48, 120–121

free-riding 77–78

Gardiner, Stephen 13, 35
geopolitics ix–x, 49–50, 71, 82–83, 162
governance 126

Hardin, Garrett 33
harm 55–60, 115–121
Hobbes, Thomas 31, 36
hunger xii, 47–48, 153–154
hypocrisy 19–21, 122–124

ignorance 72, 78–79, 81–85, 96–97
incentives xiv, 53–54, 57–60, 67–70, 72, 78–79, 109–10, 160, 181–182
inconsequentialism (see inefficacy)
inefficacy xi, 10–11, 56–57, 115–121, 138–139, 182, 183–184
integrity xiii, 5, 17–25, 121–125, 146–147, 182, 184
Intergovernmental Panel on Climate Change (IPCC) 12, 48–49

Jamieson, Dale 13
Johnson, Baylor 6–7, 34

lifestyle changes x–xii, 50, 61–63, 71–76, 89, 159–160, 163–164
"low-hanging fruit" 87, 107–8
luxury emissions 11, 140

Marx, Karl 157
McGregor, Deborah 39
migration 153
monitoring 66–67, 82–83, 161–162
moralizing xiv, 57–60, 74–75, 89, 158–162, 182–183
moral corruption 13, 16
Murphy, Patrick 22

"no regrets" climate action 60–64, 70, 85, 87–88, 151, 164–166, 183
Nolt, John 9–12, 115–121
norms (see social norms)
nuclear power 59, 81–82

obligations xi, 53–56, 58–59, 65–66, 75–78, 80, 84–87, 89, 128–131, 152, 155–156, 159–160, 165
Ostrom, Elinor 41–42, 44–45, 126

paradigmatic moral problem 13–14
Paris Agreement ix, 82–84, 162
polarization 73–76, 77–79
political action 50, 72, 76ff, 165
polycentricity 4, 41–45, 126, 165, 183
poverty xii, 47–48, 51–52, 153–154

reasons xiii, 5, 19, 31, 56, 60, 128–131, 160–162; of integrity see integrity; of relationality see relationality
relationality xiii–xv, 5, 31, 36–40, 125–128, 146–147, 184
relational ethics 31–32
role models 75, 163–164

sacrifice 150–152
self-interest 31, 125–126
Sinnott-Armstrong 9, 26–7
Smith, Adam 156–157, 176
social norms xv, 64ff, 77–80, 89, 110–13, 127, 158–162
specialization xii–xiii, 52–54, 94, 99–104, 123–124, 127–128, 131–132, 149–151, 155–158, 164–165
structural solutions xiv, 58–60, 79, 160, 184–185
subsistence emissions 11

tailored engagement 87–90, 164–166
tradeoffs 52, 71–73, 80–81, 84–85, 87–88, 154
tragedy of the commons 32–35, 125–126, 185

United Nations Framework Convention on Climate Change (UNFCCC) ix, 4, 82–84

virtue signaling 18, 71–73, 161, 186
voting 76, 186